PROCESSING AND FABRICATION TECHNOLOGY

PROCESSING AND FABRICATION TECHNOLOGY

VOLUME 3

MICHAEL G. BADER
WILBURN SMITH
ALLAN B. ISHAM
J. ALBERT ROLSTON
ARTHUR B. METZNER

Reviewing Editors
LEIF A. CARLSSON, Ph.D.   JOHN W. GILLESPIE, JR., Ph.D.

TECHNOMIC
PUBLISHING CO., INC.
LANCASTER · BASEL

Delaware Composites Design Encyclopedia—Volume 3
a TECHNOMIC publication

Published in the Western Hemisphere by
Technomic Publishing Company, Inc.
851 New Holland Avenue
Box 3535
Lancaster, Pennsylvania 17604 U.S.A.

Distributed in the Rest of the World by
Technomic Publishing AG

Printed in the United States of America
10 9 8 7 6 5 4 3 2 1

Main entry under title:
  Delaware Composites Design Encyclopedia—Volume 3/Processing and
    Fabrication Technology

A Technomic Publishing Company book
Bibliography: p.

Library of Congress Card No. 89-51098
ISBN No. 87762-701-0

C O N T E N T S

FOREWORD

The Delaware Composites Design Encyclopedia provides users with basic knowledge about the design and analysis of composite materials and structures. The six-volume indexed set is an ongoing series to which new volumes will be added as new needs arise and new knowledge is gained about this rapidly growing field of composites. Unlike many other guides on this subject, the encyclopedia emphasizes the underlying fundamental science base—documenting the high-quality research in the various disciplines that contribute to the field—in addition to engineering solutions to specific problems. It is intended for use by engineers, materials scientists, designers, and other technical personnel involved in the applications of composite materials to industrial products.

The material contained in the encyclopedia was written by international experts in the field and compiled at the University of Delaware's Center for Composite Materials (CCM). Established in 1974, CCM began its University-Industry Consortium, "Applications of Composite Materials to Industrial Products," in 1978 to meet the needs of the aerospace, automotive, electronics, and consumer products industries. Now regarded as an international leader in composites research and education, the Center continues to be supported by the Consortium as well as by the National Science Foundation, the United States Army Research Office-University Research Initiatives Program, other federal agencies, and the State of Delaware.

The first version of the *Delaware Composites Design Encyclopedia* was published in 1981 and offered as a special benefit to Consortium members. It has since grown in size and scope and is now being offered to the composites community at large for the first time. The current volumes cover the following subjects:

Volume 1— Mechanical Behavior and Properties of Composite Materials
Volume 2— Micromechanical Materials Modeling
Volume 3— Processing and Fabrication Technology
Volume 4— Failure Analysis of Composite Materials
Volume 5— Design Studies
Volume 6— Test Methods
Index to Volumes 1-6

Dr. John W. Gillespie, Jr. and Dr. Leif A. Carlsson, review editors of the *Delaware Composites Design Encyclopedia*, have incorporated the most recent references relevant to the subject matter to maintain the highest quality, up-to-date encyclopedia of this type.

Volume 3, entitled *Processing and Fabrication Technology*, is authored by international experts M. G. Bader, W. Smith, A. B. Isham, J. A. Rolston, and A. B. Metzner. This volume emphasizes the relationships among resin chemistry, rheology and properties for various composites manufacturing technologies. Molding technologies discussed include open mold laminations, compression, transfer, injection and reinforced reaction injection processes. Sections on filament winding and pultrusion are also presented. Detailed descriptions of each step in a fabrication technology provide engineers with a physical understanding of the process as well as insight into complex interactions among processing, materials, and performance. By emphasizing the fundamentals that affect material properties during manufacture, engineers and scientists will be able to select the best processing and fabrication technology that will fulfill the economic and performance requirements of the composites application.

John W. Gillespie, Jr., Ph.D.
Review Editor and Assistant
   Director for Research
Center for Composite Materials
University of Delaware
Newark, Delaware 19716 USA

SECTION 3.1

Historical Background

The first generations of composite materials were developed before their full structural potential was appreciated. Typical applications have been the use of glass reinforced plastics for the construction of small boat hulls and architectural panels. In these applications the materials have been used mainly as replacements for traditional materials such as wood. In many cases, one of the main attractions of the processes was the reduced demand for highly skilled manpower. They were, however, labor-intensive in terms of hours. Subsequently, with the introduction of the high-performance fibers such as carbon (graphite), boron, and Kevlar®,[1] a new high technology has been developed, aimed especially at the aerospace and military industries. Costs here have been considered less important than ultimate technical performance.

Extensive research and development in these fields have developed the science and technology of composite materials to the point at which their potential for general consumer industry applications has become appreciated. A primary requirement for these materials is cost-effectiveness, which has brought about a shift in emphasis from exotic products and expensive, labor-intensive fabrication techniques to the development of systems and materials suitable for large-scale, lower cost manufacturing operations. In particular, there is a need for shorter processing times and continuous operations.

Fabrication technology of composites, as well as their physical properties, is dominated by the chemistry and rheology of the matrix resins and by the type and physical form of the reinforcements. Consequently, an understanding of the availability of various matrices and reinforcements is important in any discussion of fabrication techniques. We will, therefore, now discuss currently used reinforcements and matrices separately and then examine their interactions in practical fabrication systems.

---

[1]Registered trademark of E. I. du Pont de Nemours & Co., Inc.

SECTION 3.2

Reinforcements

3.2.1 Glass Fibers

Glass fibers are manufactured by drawing molten glass from a die (or bushing) to form a strand of fibers [1]. The number of filaments in the strand corresponds to the number of openings in the bushing and is usually between 200 and 1,000. This number is termed the "strand count." The fiber diameter is determined by the temperature of the molten glass and by the drawing velocity. The preferred diameter range is 5 to 20 $\times$ 10^{-6} m (0.0002–0.0008 in), and the most common range for E-glass fibers manufactured for reinforcement purposes is 12 to 14 $\times$ 10^{-6} m (approximately 0.0005 in). Filaments of finer diameter than 10 $\times$ 10^{-6} m are less economical due to excessive filament breakage during the manufacturing process.

A necessary part of the fiber spinning operation is the incorporation of a "size," or "finish," into the strand. This size serves a number of purposes: it protects the filaments in the strand from abrasion and damage and reduces the possibility of fiber damage and degradation during subsequent handling processes such as spooling, weaving, and molding operations. Secondly, the size binds the filaments together to give a more or less coherent strand as desired. Finally, a coupling agent may be incorporated in the size to control the degree of wetting and bonding of the filaments by the matrix resin during subsequent processing. The degree of coherence of the strand and the choice of coupling agent are critical factors in determining the behavior of the glass during processing and composite fabrication; therefore, it is pertinent to consider their influence in more detail.

The size will usually be applied in the form of a water-based emulsion. This emulsion may be sprayed onto the newly drawn strand of fibers, or the fibers may be passed through a bath or over a wetted roller. The principal constituents of the size are a film-forming agent, lubricants, and coupling agents. The film formers act to bind the fibers into the strand. The choice of type and quantity of film formers determines a number of important characteristics of the strand. These characteristics are the degree of coherence and "stiffness" or "softness" of the strand, and the behavior of the strand when exposed to the matrix resin during fabrication. A heavy application of film former results in a coherent strand in which the fibers are virtually encapsulated in a continuous coating of size, whereas a light application merely gives point adhesion at intervals along the fiber and a much softer "feel" to the strand. Several alternative materials are available for use as film formers: starch, polyvinyl acetate, epoxy resins, polyesters, and polyurethane-based resins. Apart from its function in consolidating the strand, the film former influences the way in which the strand interacts with the matrix resin during fabrication of the composite. For instance, a styrene-soluble formulation will dissolve readily in unsaturated polyester resins allowing rapid "wet-out" of the strand. Another consideration is the behavior of the film former during elevated temperature processing— whether it melts, allowing the fibers in the strand to disperse, or whether it remains solid and maintains strand integrity. These considerations are very important in systems containing chopped strands where flow of the reinforcement is a feature of the molding process (as in SMC[2] and FRTP[3]), since they determine whether the fibers are individually dispersed to some degree or whether they remain bound in a strand.

In technology requiring the use of chopped strands, the film former is required to consolidate the precursor (roving) into a form suitable for efficient chopping to be carried out.

Lubricants such as waxes or soaps are also often incorporated into the size. They serve to modify the ac-

[2]SMC = Sheet Molding Compound.
[3]FRTP = Fiber-Reinforced Thermoplastic.

tion of the film former (generally softening the feel), but they also assist in reducing fiber damage during handling and especially during weaving operations.

The role of the coupling agent is controversial and complex. Its basic function is to form a chemical link between the glass surface and the resin matrix. Typically, a coupling agent will consist of a silane which contains a functional group compatible with the resin. It is generally accepted that the silane group will attach itself to the surface of the glass fiber and the functional group will react with the resin, thus forming a chemical "bridge." The exact mechanism is open to argument, but it is generally accepted that silane coupling agents perform a useful role in glass reinforced systems using unsaturated polyester, epoxy, vinyl ester, and thermoplastic resin matrices. There is often little or no measurable effect of the coupling agent on the short-term mechanical properties of the composite under dry conditions, but deterioration of properties due to water uptake does appear to be reduced in many cases when certain coupling agents are incorporated in the size.

Sometimes the size is applied in two or more stages. For instance, a finish incorporating only film former and lubricant may be applied when the fiber is drawn, and then at a later stage of processing (e.g., after weaving into cloth), the finish is removed by heat or solvent cleaning and replaced by another formulation incorporating a coupling agent. It is generally accepted that coupling agents are most effective when applied at the time of fiber drawing. However, it is sometimes necessary to use different size formulations to facilitate intermediate processes such as weaving and then to finish the cloth with a size better suited to the final molding operation.

3.2.2 Graphite/Carbon Fibers

Graphite and carbon fibers are prepared from polymer precursors, mainly polyacrylonitrile (PAN) and cellulose (rayon) but also from pitch [1,2]. They are available in "tows" of untwisted filaments, typically 2,000–10,000 individual filaments.

Fiber diameters are generally around $6–10 \times 10^{-6}$ m (about 0.0004 in), and fibers may be of circular, "dog-bone," or irregular cross section. In general, graphite fibers are more easily handled than glass so that elaborate finishes are not necessary. The as-manufactured fibers are usually subjected to a surface treatment which is designed to improve the adhesion between fiber and matrix in the final molded product. This treatment may consist of a brief oxidation by heating in air or of treatment in a solution of an oxidizing acid but, most commonly, by an anodic electrolytic treatment. In order to improve the handling characteristics of the fiber, a light size is often applied. This size usually consists of a resin that is compatible with the designated molding resin, e.g., an epoxy. If the fibers are to be woven into cloth, a heavier application of size is required. Graphite-fiber practice differs quite markedly from that of glass in that most graphite currently goes into unidirectional products, such as prepreg or pultrusion, and relatively little is used in woven or chopped form. There is, however, a move towards greater use of woven products.

3.2.3 Synthetic Organic Fibers

Polyaramid fibers, such as Kevlar, are polymeric textile fibers [1]. They have stiffness and strength intermediate between that of glass and graphite, and they have great potential for reinforcement purposes due to their excellent specific stiffness and strength. As in the case of graphite, they are available in tows or yarns of various weights and may be converted into woven cloth and chopped mat products. The fibers are tougher than those of either glass or graphite and "handle" more like a conventional textile fiber. They are, however, rather more difficult to cut, which raises some handling problems. The polyaramid fibers, as a class, appear to bond less well to conventional molding resins (epoxies and unsaturated polyesters), and the choice of sizes and coupling agents to improve this aspect of their performance is an important issue. The fibers are also weak in compression, which further limits their application to tension dominated stress fields.

3.2.4 The Form of the Reinforcement and Its Preparation for Molding

With the exception of asbestos, reinforcing fibers are first manufactured in the form of continuous filaments. The primary unit, in the case of glass, is a strand; for graphite and Kevlar it is a tow. This material must eventually be incorporated with the matrix resin into the final composite, but according to the fabrication technology, one or more intermediate conversion steps might be necessary.

The form of the reinforcement varies with the fabrication processes, which may be divided into three main categories. In the first, the fiber is placed directly onto the component being manufactured. Examples in this category are the spray-up contact molding process for GRP [where continuous glass-fiber rovings are chopped into $20-75 \times 10^{-3}$ m lengths (1–3 in) and sprayed with liquid resin onto the mold surface], the pultrusion process, and filament winding [3]. The second category comprises those processes in which the fiber is prepared in a sheet or web form which is placed in position on the mold. The resin may be incorporated before, during, or after the lay-up operation. An essential characteristic of these processes is that the fiber is placed more or less precisely in position on the molding. Most contact molding and prepreg processes come within this category. The third group comprises processes involving the preparation of a compound of fiber and resin which is required to flow during molding. Processes such as SMC, BMC(DMC), short-fiber reinforced thermoplastics, and thermosets fall in this category. With a few exceptions, these processes all use discontinuous fibers.

In glass-fiber technology, a number of individual strands (or "ends") are usually incorporated into a roving to provide a convenient aggregate of fibers for subsequent processing. A roving consists of untwisted strands, as distinct from a "yarn," which is twisted. For most reinforcement processes, rovings are preferred. The rovings are wound onto packages, typically containing 10–20 kg (20–50 lb) of fiber. Note that the strands are generally not bound in any way within the roving; i.e., no further size is applied.

The unit used to define the weight of the roving is the "tex," and the "tex count" is 10^{-6} times the mass in kilograms of one meter of roving. Thus, a single glass filament of 0.13×10^{-6} m (0.0005 in) diameter is approximately 2 decitex, a strand of 200 filaments about 70 tex, and a heavy roving of perhaps 50 strands about 3,500 tex. In the USA, a "yardage" measure is sometimes used. The "yardage count" is the number of yards of strand, roving, etc., which weighs one pound. Thus, a strand of 70 tex would have a yardage count of 7,500. A heavier roving has a *higher* tex count but a *lower* yardage. This yardage scale is illustrated in Table 3.2-1.

The packaged roving is used directly for filament winding, pultrusion, and spray lay-up processes. For other processes the rovings may be converted into other forms. The most widely used intermediate forms are sheet materials which are chopped strand mat (CSM), woven cloth, woven rovings, and prepreg [4].

CSM is manufactured by chopping rovings usually to 20–50 mm (1–2 in) lengths and allowing them to fall onto a continuously moving conveyor belt so that the chopped fibers fall in a random manner. The mat is then consolidated by the addition of a binder—either a liquid spray or powder—and then passed between rollers. Usually, heat is also applied to fuse or dry the binder, depending upon whether it is a powdered resin or a liquid emulsion. CSM is manufactured in various thicknesses and is rated on the basis of mass per unit area, i.e., kilograms/sq meter (oz/sq yard). Typical weight would be between 50 and 120×10^{-3} kg/m² (1.5–5 oz/yd²). The characteristics of CSM are determined by the weight of the mat, the strand count, the length of the chopped strands, the degree to which the fibers are bound by the size in the strand, and the degree to which the strands are bound by the mat binder.

Cloths are manufactured by weaving rovings as yarns. Simple, plain, satin, or twill weaves are usual,

Table 3.2-1. Glass-fiber terminology.

a. Fiber Diameter		U.S. Designation
10^{-6} m	inches	
5	0.00023	D
6	0.00025	DE
7	0.00028	E
9	0.00038	G
10	0.00043	H
13	0.00053	K

b. Fiber Strand Count for Basic 200 Filament Strand

Diameter (10^{-6} m)	TEX 10^{-6} kg/m	Yardage* (Yards/Pound)
5	11	45,000
6	17	30,000
7	22	22,500
9	34	15,000
10	40	12,000
13	70	7,500

Approximate conversion:
$$\text{TEX} \times \text{YARDAGE} = 500{,}000$$
i.e. $\text{TEX} = 500{,}000/\text{YARDAGE}$

*Often quoted in hundred yards per pound (h y p p.), e.g., 11 TEX − 450 h.y.p.p.

WEAVE PATTERNS FOR WOVEN REINFORCEMENT

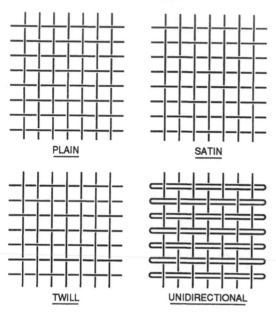

FIGURE 3.2-1.

of the glass content running in the warp direction. This weave is achieved by using a light binding weft on a heavy warp roving. These products are termed unidirectional woven rovings.

The other major group of sheet-form materials in this category are termed "prepregs." These differ from the materials previously described, in that a specially formulated resin is incorporated in the product during the molding operation. A small excess of resin is usually added, which must be bled off during molding. The principles of the resin formulation are discussed in a later section. In theory, the prepreg principle could be applied to any sheet reinforcement product, but in practice it tends to be used only for lightweight cloths and for their webs of unidirectional fibers. These products are aimed at the high technology end of the market and are suitable for molding laminates with very high fiber contents, i.e., fiber volume fractions of 0.6 to 0.7. This fiber content can be achieved only with unidirectional material. Prepreg is usually supplied in continuous rolls of tape of up to 1.0 m (39 in) width and is commonly about 0.1–0.25×10^{-3} m (0.004–0.01 in) thick. Light woven cloths are also prepregged. This prepreg is an expensive material, and consequently the process is more commonly applied to graphite, aramid, and boron fiber than to glass.

The final category of reinforcement forms includes those that are designed to flow during the molding operation, at which point the fiber is generally in a short, dispersed form. It includes Sheet Molding Compounds (SMC) and Dough Molding Compounds (DMC, also called Bulk Molding Compounds, or BMC). It also includes the short-fiber reinforcement of conventional thermoset resins (e.g., phenolics) and thermoplastics (e.g., polyamides, polycarbonates, and thermoplastic polyesters). The formulation of these materials is dominated by the polymeric component, and the systems are discussed in the appropriate specialized subsections.

SMC is a combination of chopped glass-fiber strands, mineral filler, and a resin, usually an unsaturated polyester formulation. Liquid resin, which has been premixed with the mineral filler (talc, chalk, ground limestone, etc.) and a thickening agent (typically magnesium oxide), is fed simultaneously with chopped fiber strands 10–75×10^{-3} m (0.5–3 in) long onto a temporary support film (usually polyethylene) so as to form a layer about 5×10^{-3} m (0.25 in) thick and 1–2 m (40–80 in) wide. The glass content is typically about

and the cloth is characterized by the type of weave, the tex of the warp and weft yarns (or rovings), and the weave density—the number of yarns per linear unit in the warp ("ends") and weft ("picks") directions. Picks and ends are measured as threads per cm or threads per inch in the metric and English systems. The cloth is also classified by mass per unit area, kg/m² (oz/sq ft) as in the case of the mat. The basic weave patterns are illustrated in Figure 3.2-1. A very wide range of cloth products is available ranging from less than 0.300 kg/m² (6 oz/sq ft). Cloths are usually balanced; i.e., they have approximately equal weights of fiber in the warp and weft directions. Woven rovings, however, are often produced in an unbalanced form with up to 90%

30% by volume. The mix is homogenized by passing between several sets of grooved rollers or a chain compactor and is then made up into rolls. At this stage the mix is very sticky, but after a period of maturation (2–5 days if a MgO thickener is used), the mix thickens so that it can be conveniently cut and handled for molding. If the mineral filler is omitted, more glass may be used to give a product with up to 65% glass.

The BMC/DMC products are similar in principle, except that they are bulk-mixed and usually extruded into a "rope" of 25–50 mm (1–2 in) diameter. This forms the feedstock for the molding operation. In general, shorter fibers are used for BMC/DMC than for SMC – typically 2–12 mm (0.1–0.5 in).

Fiber filled thermosets and thermoplastics are compounded by melt-mixing short chopped or milled fibers in a closed mixer, often a twin-screw extruder. The compounded product is pelletized or ground to provide the feedstock for compression, transfer, or injection molding operations. Sometimes continuous fiber rovings are fed directly into the blender, but it is a feature of all the compounding processes that the fibers are broken into very short lengths [often less than 1 mm (0.04 in)] and are fully dispersed; i.e., the strands are separated into individual fibers. In this sense these materials differ from the other discontinuous fiber materials (SMC, DMC, BMC) in which the strand integrity is usually preserved through to the finished molding.

3.2.5 References

1. HULL, D. *An Introduction to Composite Materials*. Cambridge University Press, New York, NY (1985).
2. JAIN, M. K. and A. S. Abhiraman. "Conversion of Acrylonitrile-Based Precursor Fibers to Carbon Fibers (Parts 1 and 2)," *J. Materials Science*, 22:278, 301 (1987).
3. ENGLISH, L. K. "Fabricating the Future with Composite Materials, Part I: The Basics," *Materials Engineering*, 4:15 (September 1987).
4. CHOU, T. W., R. L. McCullough and R. B. Pipes. "Composites," *Scientific American*, 255(4):192 (1986).

SECTION 3.3

Resin Systems

3.3.1 Introduction

Of the 2 billion pounds of reinforced plastics produced in the USA in 1978, about 64% were resin [1]. Matrix resins and thermoplastics thus account for most of the volume of reinforced plastics. Thermosetting resins account for about 84% of all resins and thermoplastics about 16%; their distribution by type is shown in Table 3.3-1. The thermosetting category is dominated by polyester resins,[4] which account for 88% of this category and 74% of all matrix resins. Epoxy resins are the second largest category of thermosets, accounting for about 5% of the total. All other thermosets, including phenolics, account for about 7% of the total. The popularity of polyester resins is accounted for by their ease of handling, good cured properties, and the fact that they are the lowest cost addition-curing thermoset available.

3.3.2 Polyester Resins

The term *polyester resin* is used to describe a class of thermosetting resins consisting of an unsaturated backbone dissolved in a reactive monomer. The most common backbone polymer consists of a saturated acid, an unsaturated acid, and one or more glycols; the most common reactive monomer is styrene (see Figure 3.3-1).

Polyester resins are commonly categorized as "ortho" or "iso" resins depending on the nature of the saturated acid portion of the backbone polymer. Orthophthalic (ortho) resins use orthophthalic acid as the saturated

Editor's Note: Following section 3.3, the reader may wish to consider surveying section 3.13 on flow behavior ahead of the more applied processing sections 3.4 to 3.12.

[4]The term *polyester resin* is used throughout this section to mean unsaturated polyester resin. Saturated polyester resins are referred to as *thermoplastic polyesters*.

acid portion of the backbone polymer. They are the least expensive and most common type, accounting for about 80% of all polyester resins used in the USA. Their largest market is contact molding (e.g., marine applications). Isophthalic (iso) resins are those having isophthalic acid as the saturated acid portion of the backbone polymer. Their cost is slightly higher (~ 5–10%) than for ortho resins, and they account for about 15% of all polyester resins. Iso resin use is concentrated in matched-die molding resins, corrosion resistant applications, and gel coats.

Manufacture

The polyester backbone polymer is synthesized by condensation polymerization, batch or continuous, of the acid/glycol combination needed for a selected formulation. A typical reaction for the preparation of an orthophthalic polyester resin is shown in Figure 3.3-2. The monomer mixture is cooked until the desired molecular weight has been reached, diluted with styrene monomer and a free radical polymerization inhibitor, and cooled. After various additives have been blended in (e.g., promoters, thixotropes, UV inhibitors, and wax), the resin is ready for packaging. Molecular weight of the backbone polymer is controlled by the manufacturer and varied to suit his particular formulation. Molecular weights are usually in the 1,000–4,000 range [2].

The ingredients in the reaction illustrated above are those most commonly used in polyester resins. Phthalic and maleic acids (used as the anhydrides) account, respectively, for about 77% and 97% of the saturated and unsaturated acids used in the USA (in 1976). Likewise, propylene glycol is by far the most commonly used glycol, accounting for about 68% of total glycol consumption [3]. These and other ingredients used in the polyester backbone are summarized in Table 3.3-2 by their function.

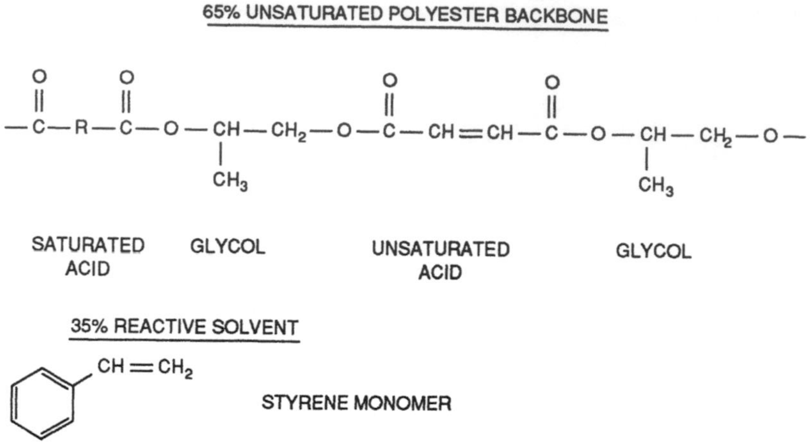

65% UNSATURATED POLYESTER BACKBONE

$$-\overset{\overset{O}{\|}}{C}-R-\overset{\overset{O}{\|}}{C}-O-\underset{\underset{CH_3}{|}}{CH}-CH_2-O-\overset{\overset{O}{\|}}{C}-CH=CH-\overset{\overset{O}{\|}}{C}-O-\underset{\underset{CH_3}{|}}{CH}-CH_2-O-$$

| SATURATED ACID | GLYCOL | UNSATURATED ACID | GLYCOL |

35% REACTIVE SOLVENT

CH=CH$_2$

STYRENE MONOMER

FIGURE 3.3-1. Typical composition of a polyester resin.

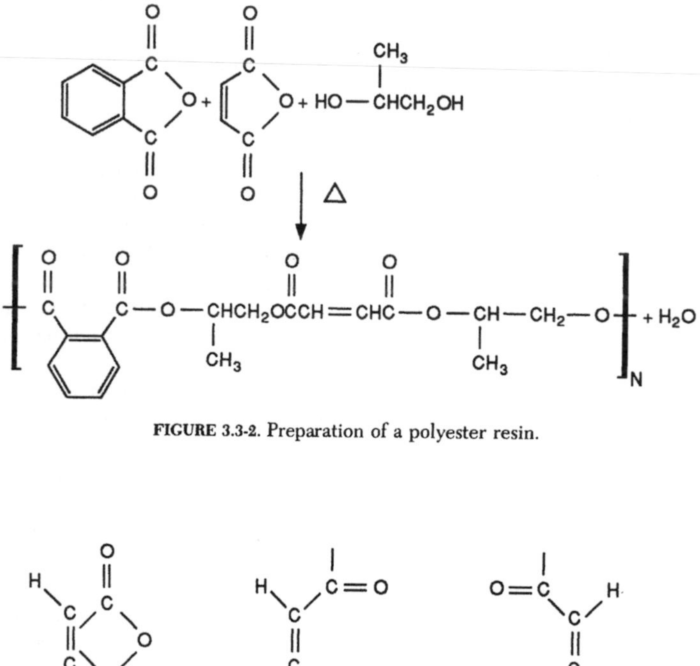

FIGURE 3.3-2. Preparation of a polyester resin.

| MALEIC ANHYDRIDE | MALEATE GROUP (CIS) | FUMARATE GROUP (TRANS) |

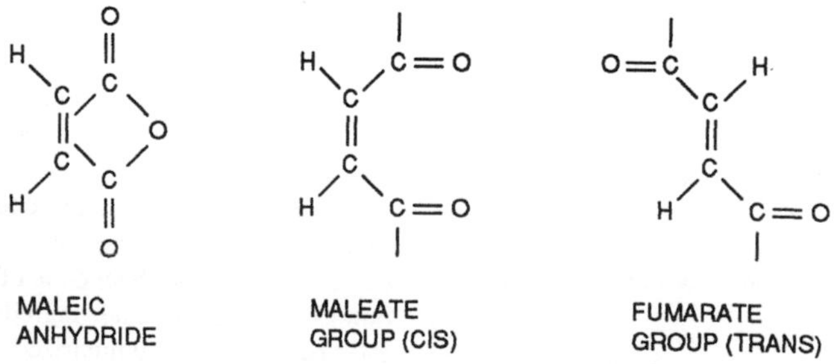

FIGURE 3.3-3. Isomerization of maleate to fumarate.

Table 3.3-1. *Sales of resins used in reinforced plastics (1978) [1].*

	10^6 lb	%
Thermosets		
Polyester	951	88
Epoxy	55	5
Other (incl. phenolics)	75	7
Total Thermosets	1,081	100
Thermoplastics		
Polypropylene	68	33
TP polyester	44	21
Nylon	38	18
ABS, SAN, Styrenics	28	14
Polycarbonate	10	5
Other	18	9
Total Thermoplastics	206	100
Grand Total (TS + TP)	1,287	–

Table 3.3-2. *Ingredients used in polyester backbone resin by function.*

	Ingredient	Function
Unsaturated acids and/or anhydrides	maleic anhydride	provide cure site
	fumaric acid	provide best cure site (maleic isomerizes to fumaric)
Saturated acids and/or anhydrides	phthalic anhydride	low cost, hard, good balance of properties,
	isophthalic acid	improved strength and chemical resistance
	adipic acid and homologs	flexibility, toughness, resiliency
	halogenated acids/ anhydrides	flame retardance
Glycols	propylene glycol	good balance of properties at lowest cost
	diethylene glycol dipropylene glycol	flexibility, toughness
	bisphenol A/PG adduct	chemical resistance, high HDT
	neopentyl glycol	chemical resistance, toughness

Effect of Backbone Structure on Properties

Effect on Reactivity. The most important structural features of the backbone polymer affecting reactivity are the total amount of unsaturation and the ratio of maleic to fumaric unsaturation. The amount of unsaturation is expressed as the mole % of total acid that is unsaturated (either as maleic or fumaric) and may vary from about 25% for a low reactivity resin to 100% for a very high reactivity resin [4]. The ratio of maleic to fumaric (cis to trans) unsaturation in the backbone polyester is usually determined by the degree to which maleic isomerizes to fumaric during preparation of the polymer rather than by the use of fumaric acid itself, since maleic anhydride is the source of most unsaturation. The degree of maleate to fumarate isomerization depends not only on the esterification conditions but also on the glycol used [5]. Isomerization of maleate to fumarate (Figure 3.3-3) is desirable because of the more favorable reactivity ratio of fumarate with styrene [2], which leads to greater conversion of backbone fumarate (vs. maleate) double bonds in the subsequent curing reaction.

Both the effect of total unsaturation and maleate/fumarate ratio on gel time and peak exotherm temperature are illustrated in Table 3.3-3. Although resins 1 and 2 have the same total mole % unsaturation, No. 2 is considerably more reactive (shorter gel time and higher peak exotherm) because all of its unsaturation is present as fumarate. Compositions 2 and 3 have all-fumarate unsaturation, but No. 3 with 100 mole % unsaturation

is much more reactive than No. 2 with only 50 mole % unsaturation.

Effect on Cured Properties. The effect on cured resin properties of varying both the type of saturated acid and the total unsaturation is shown in Table 3.3-4 for four compositions. Compositions 1 and 2 show the effect of the isophthalic vs. orthophthalic acid as the saturated acid component at constant maleic acid content. The major property improvements resulting from the use of

Table 3.3-3. *Effect of backbone structure on reactivity [6].*

Resin No.	Composition of Backbone Polymer (moles)	SPI Gel Time (minutes)	Peak Exotherm Temperature, °F
1.	4.5 MA, 0.5 FA 5.0 PA, 10 PG	6.1	227
2.	5.0 FA, 5.0 PA, 10 PG	3.6	417
3.	10 FA, 10 PG	1.2	469

MA = maleic acid.
FA = fumaric acid.
PA = phthalic acid.
PG = propylene glycol.

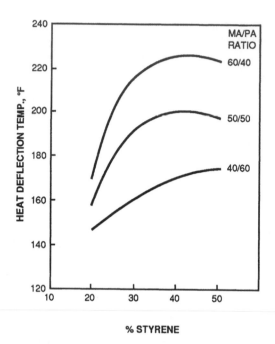

FIGURE 3.3-4. Effect of styrene concentration on heat deflection temperature of propylene glycol/maleate/phthalate resins [plotted from data of M. C. Slone, *SPE Journal*, 16: 1123 (1960)].

isophthalic acid are increased heat deflection temperature (HDT), increased tensile and flex strengths, and slightly higher ultimate elongation. Increasing the amount of isophthalic acid at the expense of maleic (i.e., reducing total unsaturation, composition 1 vs. 3) lowers HDT significantly due to reduced cross-link density but improves tensile strength and elongation. On the other hand, increasing the maleic acid content gives higher total unsaturation leading to higher cross-link density (composition 4 vs. 1 and 3) with resulting high HDT but lower tensile strength and elongation.

Typical cast resin elongations, as shown in Table 3.3-4, are low for both isophthalic and orthophthalic resins, especially at higher levels of unsaturation. The usual route to more "flexible" or "resilient" resins is to modify the backbone polymer by the use of an aliphatic acid and/or glycol segment such as adipic acid or diethylene glycol (Table 3.3-2).

Cross-linking Monomers

One major function of the reactive monomer is to dissolve the viscous backbone polyester resin to give a relatively low viscosity (350 cps to ~4,000–5,000 cps) solution which can be used to impregnate fibers or fillers prior to cure. A second major requirement is that

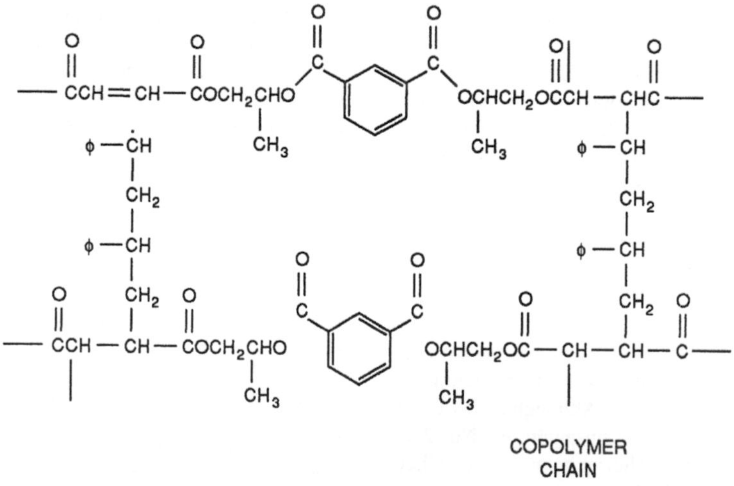

FIGURE 3.3-5. Curing polyester resins.

the monomer copolymerize well with the backbone unsaturation during the curing reaction. A number of monomers meet these two basic requirements, but by far the most widely used is styrene, which accounts for more than 80% of all monomer consumption in polyester resins; the next most important monomers are methyl methacrylate and vinyl toluene [3]. These and other monomers are listed in Table 3.3-5 with major characteristics. Methyl methacrylate is used in applications where improved weatherability is desired (e.g., in construction panels). Due to its poor copolymerization characteristics with the polyester backbone, methyl methacrylate is used not as the sole monomer but as a mixture with styrene; optimum panel weatherability occurs at about 50/50 styrene/methyl methacrylate [8].

The effect of the monomer type and concentration on various cured resin properties is complex because of the number of variables. For more information and leading references, see reference [2]. The effect of varying styrene levels on the heat deflection temperature (HDT) is shown in Figure 3.3-4; the HDT increases with increasing styrene concentration at all levels of unsaturation but reaches a maximum in those resins having high levels of unsaturation.

Table 3.3-4. Variation of polyester resin properties with structure of backbone polymer [7].

	1	2	3	4
Materials				
Isophthalic acid	1.0		3.0	1.0
Phthalic anhydride		1.0		
Propylene glycol	2.2	2.2	4.4	3.3
Maleic anhydride	1.0	1.0	1.0	2.0
Styrene, %	45	45	45	45
Maleic, mole % of total acid	50	50	25	66
Uncured Resin				
Acid Number	16	18	10	17
Gel time, minutes	2.0	2.25	3.75	1.75
Cured Resin				
Heat deflection temp., °C	110	95	85	125
Tensile strength, 10^3 psi	9	5	12	6
Tensile modulus, 10^6 psi	500	500	500	500
Elongation, %	1.8	1.4	2.5	1.3
Flexural strength, 10^3 psi	19	16	19	14
Flexural modulus, 10^6 psi	570	600	520	580
Relative water permeation rate				
23°C	0.18	0.13	0.05	0.22
65°C	32.8	31.8	22.9	46.6

Table 3.3-5. Cross-linking monomers used with polyesters.

Monomer	B.P. °C	Characteristics
Styrene	145	low cost, good reactivity, high shrinkage
Chlorostyrene	188	faster cures, high exotherms, lower shrinkage
Vinyl toluene	172	
α-methyl styrene	165	lower exotherm, slower cures
Methyl methacrylate	100	improved weatherability
Diallyl phthalate	160	low volatility; useful in prepregs
Triallyl cyanurate (TAC)	M.P. 27°	good high temperature performance (up to 200°C)
T-butyl styrene	219	low volatility

Curing Polyester Resins

General. Polyester resins cure by a free radical copolymerization of backbone double bonds (fumarate or maleate) with styrene monomer (or other reactive diluent). The curing reaction produces a cross-linked polymer in which the polyester chains are bridged by one or more styrene units and become incorporated in a long copolymer chain, as shown in Figure 3.3-5. The length of the copolymer chains in the cross-linked resin is several times longer than the length of the condensation polyester chains ($MW = 8,000–14,000$ vs. $1,000–3,000$) [2]. In general, the objective in the curing step is to maximize the conversion of backbone double bonds, and, to do this, homopolymerization of the monomeric species must be kept to a minimum. Optimum physical properties of the cured resin are achieved when the molar ratio of styrene to fumarate unsaturation is about 2:1 [2]. At this level, conversion of both monomer and backbone double bonds is high.

The curing reaction is initiated by a free radical source which is typically an organic peroxide chosen to provide radicals at elevated temperature by thermal decomposition of the peroxide or at room temperature by "accelerated" or "promoted" decomposition of the peroxide. After addition of the initiator there is an induction period during which inhibitor is consumed, followed by gelation and the generation of a large exo-

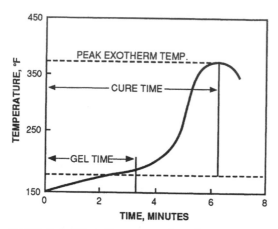

FIGURE 3.3-6. SPI exotherm curve for polyester resin at elevated temperature.

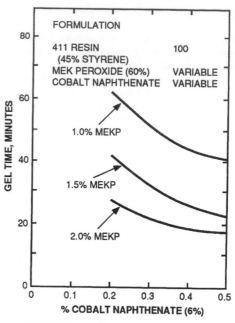

FIGURE 3.3-8. Gel time vs. percent accelerator or vinyl ester resin (data from Dow Chemical Co.).

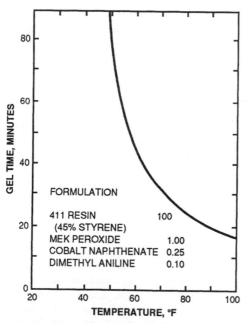

FIGURE 3.3-7. Gel time vs. temperature for a vinyl ester resin (data from Dow Chemical Co.).

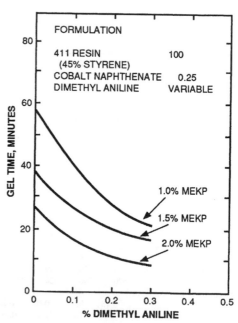

FIGURE 3.3-9. Gel time vs. percent DMA for vinyl ester resin (data from Dow Chemical Co.).

therm. Figure 3.3-6 shows this exotherm for a heat-cured resin in an SPI test used to compare resins or initiators under a standard set of conditions [9]. The magnitude of the peak exotherm temperature depends on the total unsaturation present and, therefore, increases with styrene content and with maleate/fumarate content [2].

Initiators and Promoters. One of the most important reasons for the popularity of unsaturated polyester resins in contact molding applications is their ability to cure rapidly and satisfactorily at room temperature. Initiation of the room temperature curing reaction is accomplished by causing the decomposition of an otherwise stable peroxide to occur at low temperature. The most common peroxide for this type of cure is methyl ethyl ketone peroxide (MEKP), which alone is thermally stable to fairly high temperatures (half-life of 12–15 hours at 100°C) but which decomposes readily at room temperature in the presence of small amounts of transition metal ions. Commercially, the most important source of these ions is cobalt naphthenate, which is typically used at about 0.01% of the resin weight (0.1–0.3% of a 6% solution of cobalt naphthenate in dimethyl phthalate) to give a "promoted" or "pre-accelerated" resin which can then be cured by the addition of MEKP at about the 1% level. Because of reactions of cobalt ions with the radicals which are generated, there is an optimum accelerator level for most efficient curing [10]. Small amounts of tertiary amines are often added along with cobalt naphthenate to give a "doubly" promoted system.

The second most widely used room temperature curing peroxide system is benzoyl peroxide (BPO), a solid diacyl peroxide which is available as a paste in several concentrations. Unlike MEKP, BPO is not promoted by cobalt but does respond well to tertiary amines (e.g., dimethyl aniline) [10]. Because it is a solid or paste, BPO is not used in spray-up equipment where a liquid initiator is mixed with resin as it leaves the gun.

For a given polyester resin, the gel time, which determines how much working time is available to the user, is affected by the initiator level, the accelerator level, the presence of a second accelerator, and the ambient temperature. The influence of each of these variables on the gel time is shown in Figures 3.3-7, 3.3-8, and 3.3-9 for a typical vinyl ester resin. In practice, a large portion of polyester resins used in contact molding is supplied as pre-accelerated or promoted resins, which usually means that the manufacturer has added everything

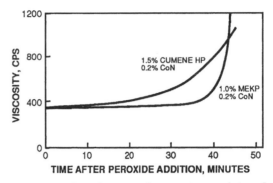

FIGURE 3.3-10. Typical viscosity changes prior to gelation of Derakane 470-36 (data from Dow Chemical Co.).

necessary for cure except the initiator; in these cases, the level of initiator and the ambient temperature are the user's major controls over gel time. In some cases, the type of initiator may be varied if a different viscosity-time behavior is desired; these cases are illustrated in Figure 3.3-10, where the same resin, cured with cumene hydroperoxide or with MEKP, gives different viscosity vs. time profiles and different peak exotherm temperatures. After addition of initiator, polyester resins typically show rather flat viscosity-time profiles until the gel point is reached, at which time viscosity increases extremely fast.

The usual method of monitoring the cure of polyester resins is to follow the Barcol hardness. Well-cured resins should have Barcol hardness values between 30 and 60 depending on resin type, reinforcement or filler type, and whether a room temperature system has been postcured. The manufacturer's recommendation should be followed.

Two factors prevent full cure at the surface of polyester laminates exposed to the atmosphere: inhibition of the radical curing reaction by oxygen and loss of styrene monomer [11]. Both of these problems can be minimized by the use of a small amount of wax dissolved in the resin. During cure, the wax exudes to the surface, where it forms a film that keeps oxygen out and styrene in. Since the wax layer acts as a mold release, it must be removed if subsequent layers are to be added.

Polyester resins used in such processes as compression and injection molding, pultrusion, and continuous laminating are cured at elevated temperatures, usually in the 250–350°F range. Rapid cure at the processing temperature is of major importance; it is a function of

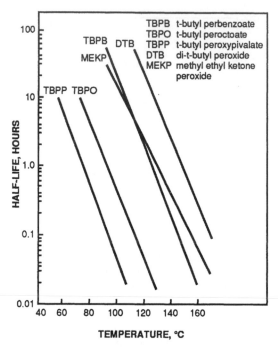

FIGURE 3.3-11. Half-life vs. temperature for organic peroxides.

part thickness, resin type, initiator, and temperature. In contrast to room temperature curing systems where gelation is usually measured in minutes following addition of peroxide, many press molding compounds must remain stable at room temperature for days or weeks after the initiator has been added; the initiator must, therefore, be stable in the presence of all components of the mix. The major initiator types used for this purpose are peresters (e.g., t-butyl perbenzoate and t-butyl peroctoate) [12]. Typical half-life vs. temperature plots are shown in Figure 3.3-11. [Half-life is the time required for one-half of a given amount of peroxide to decompose at a given temperature (in seven half-lives, >99% of the peroxide has decomposed).] Although these half-life data are determined in solvents, they are useful for estimating cure times in press moldings [10].

High-Performance Polyester Resins

A number of "high-performance" polyester resins are used primarily because of their corrosion resistance and high heat deflection temperatures. The two most important types are bisphenol A/fumarate resins and vinyl ester resins. The bisphenol A/fumarate resins (typ-

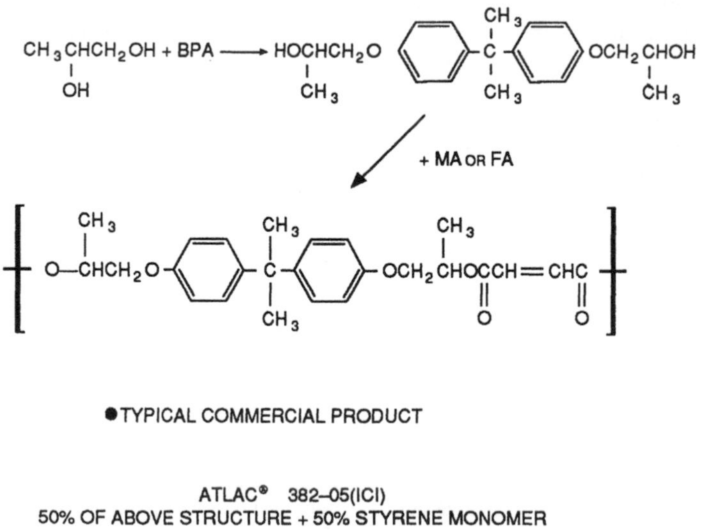

FIGURE 3.3-12. Bisphenol A/fumarate resins.

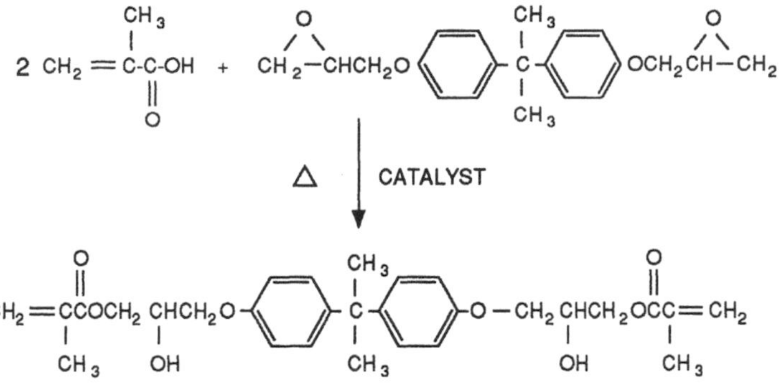

FIGURE 3.3-13. Preparation of vinyl ester resins.

ified by Atlac 382-05 from ICI) were the first widely used premium corrosion resistant resins and are still widely used in both the USA and Europe [13]. These resins are styrene solutions of polyesters from a bisphenol A/propylene glycol adduct and fumaric acid with the structure shown in Figure 3.3-12. Vinyl ester resins are acrylic esters of epoxy resins dissolved in styrene monomer. They differ from typical polyester resins in having only terminal unsaturation in the backbone, pendant hydroxyl groups, and no carboxyl or hydroxyl end groups. They are synthesized from an unsaturated carboxylic acid (usually methacrylic) and an epoxy resin whose structure is varied depending on the properties needed in the final resin [14]. The most important commercially available vinyl ester is prepared from methacrylic acid and a diglycidyl ether of bisphenol A as shown in Figure 3.3-13. Fire retardant and high temperature grades are based on tetrabromobisphenol A epoxies and on epoxy novolac resins, respectively.

Table 3.3-6. Properties of commercial polyester resins.

Resin Type and Manufacturer's Description	Trade Name and Designation		HDT °C 264 psi	Tensile			Flexural		Barcol Hardness	Resin Viscosity CPS
				Str. ksi	Elong. %	Mod. ksi	Str. ksi	Mod. ksi		
Ortho, Hand Lay-up	Hatco	MR480	75	9	–	–	20	600	50	375
	Polylite	33-072	–	8–10	–	–	12–15	575	45–50	375
ISO, Laminating	Co-Rezyn	9501	93	8.0	1.5	640	15.0	530	–	–
ISO, Corr. Resist., Molding	Hatco	MR14042	92	10.6	2.5	–	17.6	530	45	650
ISO, Press Molding	Aropol	7241	126	9.1	1.5	–	23.4	560	45	500
ISO, SMC-BMC	Co-Rezyn	349	70	11.0	4.0	500	18.0	450	35	3,900
ISO, Fil. Wind., Resilient	Stypol	40-2586	55	10.7	7.8	250	18.7	540	35	550
ISO, Flex., Blending Resin	Stypol	40-2092	–	2.5	90	–	–	–	–	800
ISO, DAP, Resilient, Premix	Stypol	40-2057	80	7.2	3.8	270	17.4	490	49	23,000

Trade Name	Company
Hatco	USS Chemicals Co.
Polylite	Reichhold Chemical Co.
Co-Rezyn	Interplastic Corp.
Aropol	Ashland Chemical Co.
Stypol	Freeman Chemical Corp.

Table 3.3-7. Properties of vinyl ester resin castings (manufacturers' data).

	Derakane® (DOW)		Epocryl® (Shell)		Co-Rezyn® (Interplastic)		Hetron® (Ashland) 902	Atlac® (ICI) 580-05
	411–45	470–45	322	480	VE-8300	VE-8710		
Heat Deflection Temp., °C	101	138	99	121	105	118	118	118
Tensile								
Modulus, 10^3 psi	490	510	450	470	470	500	–	–
Strength, 10^3 psi	11.8	11.0	11.4	12.2	11.6	11.0	9.2	13.1
Elongation, %	5.0	4.0	5.5	5.3	5.0	3.5	2.2	4.2
Compressive								
Modulus, 10^3 psi	350	320	380	454	–	–	–	–
Strength, 10^3 psi	16.6	18.5	16.5	18.9	–	–	–	–
Def. at Yield, %	6.6	11.9	6.6	8.0	–	–	–	–
Flexural								
Modulus, 10^3 psi	450	550	465	470	450	470	520	490
Strength, 10^3 psi	18.0	20.0	19.4	17.0	19.4	18.0	14.3	22.6
Barcol Hardness	35	40	35	35	37	45	–	40
Specific Gravity	1.12	1.15	1.12	–	1.12	1.13	–	–

®Trade Names: See Table 3.3-9.

Cured Resin Properties

A complete set of mechanical properties for neat resins is seldom available unless measured by the user. The manufacturer's data usually include flexural and tensile properties plus heat deflection temperature and the Barcol hardness to be expected for well-cured specimens. Table 3.3-6 provides mechanical property data for several commercially available resins from various suppliers. Note that flexural modulus is usually in the 450–600 ksi range, tensile strength in the 8–11 ksi range, and elongation in the 1.5–4.0% range except for some "resilient" grades and highly flexibilized resins used for blending with rigid types. Heat deflection temperatures vary from 55°C for a "resilient" grade to 126°C for rigid press molding grades. Table 3.3-7 shows similar data for several vinyl ester and bisphenol A/fumarate (BPA/F) resins. Compared to the ortho and iso resins, the vinyl ester and BPA/F resins show the following:

• higher heat deflection temperatures (100–138°C)
• higher tensile strength (9–12 ksi) and elongation (2.2–5.5%)
• about the same range of flexural modulus values
• lower Barcol hardness

Chemical Resistance. Polyester resins at room temperature show generally good short-term resistance to common acids, bases, and solvents. At elevated temperatures and for long exposure times, the chemical resistance depends on the structure of the resin and, for most environments, follows the order [13]:

$$\left.\begin{array}{c}\text{bisphenol A/fumarate} \\ \\ \text{vinyl esters}\end{array}\right\} > \text{isophthalics} > \text{orthophthalics}$$

This order is found in Figure 3.3-14, where flex strength retention vs. exposure time is shown for identical laminates from three resins exposed to boiling water for up

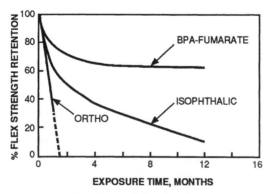

FIGURE 3.3-14. Degradation of glass CSM laminates in water at 100°C for different polyester resin matrices [15].

to one year [15]. The orthophthalic resin is obviously not suited for exposure to boiling water and is typically used only at about room temperature in water. In boat hull applications, where ortho resins are the predominant matrix material, completely satisfactory performance has been obtained after long service in water [16]. See also Figure 3.3-15.

The mechanism of attack of water and other chemicals on polyester/glass laminates has been the subject of many studies [17,18,19]. The most important degradation mechanisms include hydrolysis of the resin itself, destruction of the fiber-resin interface, and physical damage from osmotic pressure effects or from swelling forces [20]. The quality of the fiber-resin interface is a strong function of the sizing agent [21].

Thermal Properties. The most commonly quoted thermal property is the heat deflection temperature (HDT), in which the value is determined by the ability of the polyester network to resist thermally activated molecular motions. The HDT is a short-term property and is not a measure of long-term thermal stability. Polyester resin HDT values of 50–75°C are common for hand lay-up orthophthalic resins, whereas values as high as 138°C have been reported for high heat vinyl esters (see Table 3.3-7). The HDT is a measure of the thermal sensitivity of the resin, which is shown in Figure 3.3-16, where flex strength vs. temperature is plotted for a standard vinyl ester (Derakane 411-45, HDT = 102°C) and high heat vinyl ester (Derakane 470, HDT = 138°C) laminates. Note that above the HDT, flex strength decreases rapidly. Thermal expansion data for a number of resin types are shown in Table 3.3-8. Included are isophthalic, brominated isophthalic, bisphenol A/fumarate, and vinyl ester types. The coefficient of linear thermal expansion values are in the 75–85 × 10⁻⁶°C range for all but a resilient grade. Lowest values are found for highly cross-linked resins with high HDT.

Flammability and Smoke. Fire retardant polyester resins account for about 3% of total polyester resin consumption. They are used primarily in corrosion resistant equipment (ducts, fume hoods, etc.), molded electrical parts, panels (construction, ballistic), and some marine applications. Fire retardancy is obtained by

- use of a halogenated resin alone or with a synergist such as antimony trioxide
- use of fire retardant additives, usually alumina trihydrate

Smoke density (in the absence of fire retardants) has

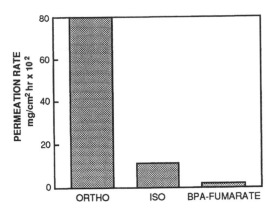

FIGURE 3.3-15. Water permeation rates for glass CSM laminates in water at 100°C for different polyester resin matrices [15].

been shown [25] to depend on the polyester resin structure and the cross-linking monomer used—styrene as the monomer gave the highest smoke level and methyl methacrylate the lowest. Of several glycols typically used in the backbone, propylene gave the highest smoke density; for the same saturated/unsaturated acid ratio, orthophthalic based resins gave higher smoke levels than isophthalics. Increasing the unsaturated acid content for isophthalic/maleic resins decreased smoke density. In-

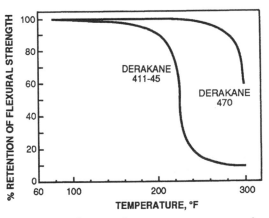

FIGURE 3.3-16. Flex strength retention vs. temperature for Derakane vinyl ester resin laminates [22].

Table 3.3-8. Linear thermal expansion of resin castings [23].

	Average Coeff. Linear Thermal Expansion 10^{-6} in/in/°C	Temp. Range °C	HDT °C 264 psi	Tensile		Flexural	
				Strength ksi	Elong. %	Mod. ksi	Str. ksi
Isophthalic (Dion 6631T)	78	20–94	107	9.3	2.4	520	16.6
Brominated Isophthalic (Dion 6693FR)	79	20–94	107	10.4	2.0	560	13.4
Modified BPA/Fumarate (Dion 6694)	77	20–94	135	8.2	2.4	490	14.6
Resilient BPA/Fumarate (Dion 7000)	106	20–94	110	9.1	4.6	420	15.2
Standard Vinyl Ester* (Epocryl 322)	84	–	99	11.0	5.5	465	17.0
High-Heat Vinyl Ester* (Epocryl 480)	75	–	121	12.2	5.3	470	17.0

*Data from reference [24].

terestingly, the most common polyesters are made with the worst smoke contributors in each category.

In the presence of fire retardant additives such as antimony trioxide, smoke levels increase, but the presence of alumina trihydrate reduces smoke evolution [26,27]. Bromine is the most common halogen used and is typically incorporated into the polymer backbone as a brominated anhydride (e.g., tetrabromophthalic) or glycol (e.g., dibromoneopentyl glycol) to give a resin with 10–25% bromine by weight. Flame spread ratings of 15 in the E-84 tunnel test have been reported for a resin with 15% bromine content and 3% antimony trioxide [28,29].

Whether the fire retardant is chemically combined with the backbone polymer or present only as an addi-

tive, it should not degrade the mechanical or electrical properties or the chemical resistance of the matrix. Resins now available commercially meet these requirements and give flame spread ratings of 25 or less with viscosities suitable for contact molding operations [30,31].

Commercially Available Polyester Formulations

Polyester resins are available in an enormous variety of formulations by a large number of suppliers. Some of the major suppliers and their trade names are listed in Table 3.3-9. Resins from these suppliers are usually optimized for use in a specific process (e.g., hand lay-up, press molding, pultrusion) and/or in a specific end use (e.g., corrosion resistance, electrical, panel manufacturing). Selection of a resin from any category involves consideration of several variables common to all systems:

Table 3.3-9. Major USA suppliers of polyester resins (partial list).

Company	Trade Names
Reichhold Chemicals, Inc.	Polylite
USS Chemicals Co.	Hatco, Laminac
Koppers Co.	Koplac, Dion
Ashland Chemical Co.	Aropol, Hetron
Owens-Corning Fiberglas Corp.	—
Freeman Chemical Corp.	Stypol, Acpot, Nupol
Interplastic Corp.	CoRezyn
ICI Americas, Inc.	Atlac
Dow Chemical Co.	Derakane
Shell	Epocryl

- Viscosity affects speed and degree of wet-out of reinforcements. It varies from 350–500 cps for contact molding systems to 5,000–6,000 cps for some press molding resins.
- Reactivity affects cure time and peak exotherm temperatures. High exotherm is needed for thin section curing at R.T. and low exotherm for thick sections.
- Resiliency—"Rigid" resins have low elongation and high modulus and Heat Deflection Temperature

(HDT); resilient or flexible grades have higher elongations at the expense of modulus and HDT.

- HDT can vary from 50°C for some orthophthalics to 130°C for press molding isophthalics. It is lowered by flexibilizers and raised by higher cross-link density.
- Presence of additives such as promoters, thixotropes, wax UV inhibitors, low profile additives, etc., is seen.

Specific end uses may impose other requirements on the matrix. For example, choice of a resin for a chemical tank requires consideration of all of the above, plus resistance of the resin to the particular chemical involved. Likewise, electrical properties, weatherability, long-term thermal stability, fire retardance, and thickenability (for SMC) are important considerations.

3.3.3 Epoxy Resins

Introduction

Epoxies are resins of widely different structures characterized by the presence of a 1,2-epoxide group:

$$\underset{\underset{\displaystyle O}{\diagup\diagdown}}{-\!\!\overset{\displaystyle |}{C}\!-\!\overset{\displaystyle |}{C}\!-}$$

This epoxide group is a 3-membered cyclic ether which is reactive with a wide variety of reagents. The epoxide group is usually present as a glycidyl ether or amine or as part of an aliphatic ring system:

GLYCIDYL ETHER $\quad CH_2\!\!-\!CHCH_2\!-\!O\!-\!Ar\ (R)$

GLYCIDYL AMINE $\quad \left(CH_2\!\!-\!CHCH_2\right)_{\!2}\!\!-\!N\!-\!Ar$

CYCLOALIPHATIC

The glycidyl ethers and amines account for most of the epoxy resins used in composites and are our major topic; cycloaliphatic resins are used in many electrical applications but are not used widely in laminates.

Epoxy resins are characterized by good mechanical strength, chemical resistance, and electrical properties. They form the second largest category of matrix resins used in composites (after polyesters), accounting for about 5% of total thermoset resin sales in reinforced plastics. Overall, the largest use for epoxies is in coatings (45% of total epoxy mass used in 1978), followed by composites (17%) and other applications in casting and adhesives [1]. Epoxy use in composites is dominated by prepregs, printed circuit board, and filament winding applications, which account for 80% of composite usage.

Commercial Types

Diglycidyl Ethers of Bisphenol A (DGEBA). The diglycidyl ethers of bisphenol A (DGEBA) are the most widely used type, accounting for about 90% of all epoxies. They are synthesized from epichlorohydrin and bisphenol A in the presence of sodium hydroxide, as shown in Figure 3.3-17. Excess epichlorohydrin is used in the "taffy" process to minimize the formation of higher molecular weight species by further reaction of DGEBA with bisphenol A, which results in liquid resins where the $n = 0$ species predominates. Higher molecular weight resins can be made by the "advancement" process, in which DGEBA is allowed to react with controlled amounts of bisphenol A to form products with $n = 2, 4, 6$, etc. [32]. Solid resins result when n is about 2, as shown in Table 3.3-10, where commercial examples of resins made by these routes are provided. The "workhorse" liquid resins shown in Table 3.3-10 have values of $n = \sim 0.1$–0.2 and viscosities in the 6,000–16,000 cps range; they are supplied by all manufacturers in nominally equivalent grades. Note that, as the value of n increases, the weight per epoxide (epoxide equivalent weight) increases because the number of epoxide groups per molecule remains constant at 2. The maximum cross-link density achievable by reaction of epoxide groups, therefore, decreases significantly, and as a result, resins with higher values of n are not used as matrix resins. These resins are widely used in coating applications, where they are cross-linked through both the epoxy and the hydroxyl functionality.

Glycidyl ethers of novolac resins. These resins form the second major category of glycidyl ethers; they are prepared from the novolac resin and epichlorohydrin. They offer higher functionality than DGEBA resins,

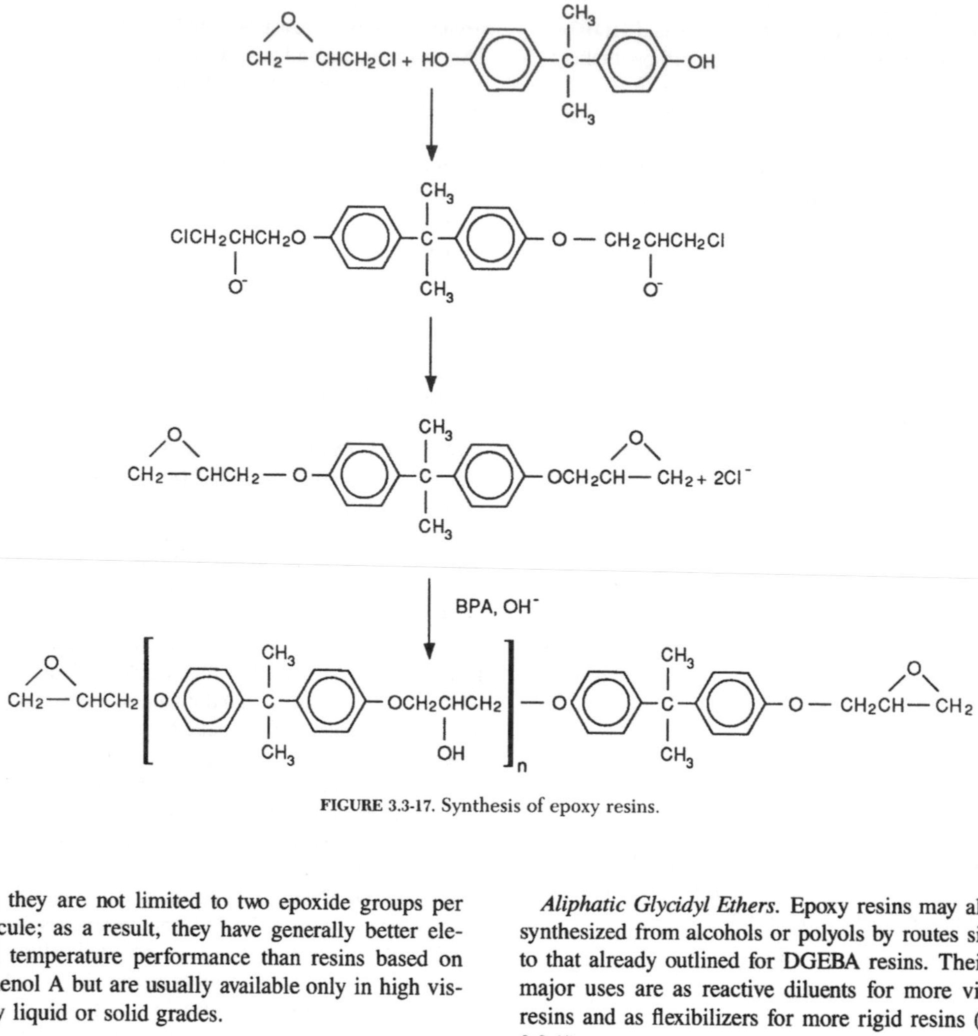

FIGURE 3.3-17. Synthesis of epoxy resins.

since they are not limited to two epoxide groups per molecule; as a result, they have generally better elevated temperature performance than resins based on bisphenol A but are usually available only in high viscosity liquid or solid grades.

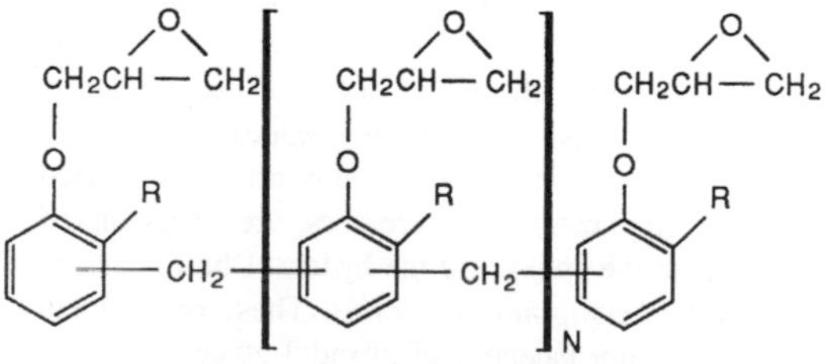

Aliphatic Glycidyl Ethers. Epoxy resins may also be synthesized from alcohols or polyols by routes similar to that already outlined for DGEBA resins. Their two major uses are as reactive diluents for more viscous resins and as flexibilizers for more rigid resins (Table 3.3-11).

Glycidyl Hydantoins. These relatively new resins, de-

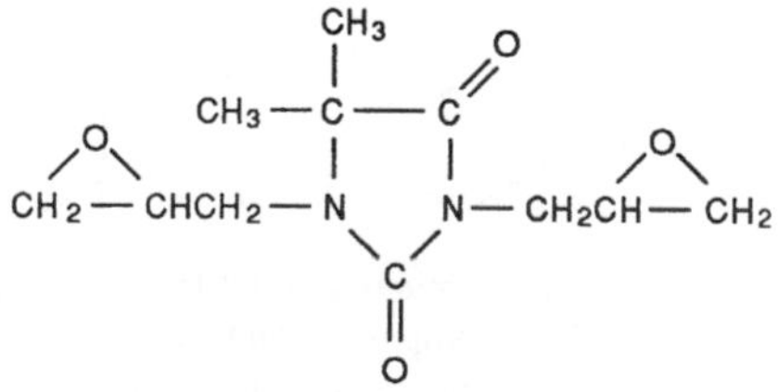

R = H EPOXY PHENOL NOVOLAC
R = CH₃ EPOXY CRESOL NOVOLAC

CIBA-GEIGY XB-2793

Table 3.3-10. *Commercial epoxy resins based on bisphenol A.*

n (approx.)	Wt. per Epoxide	Viscosity or M.P.	Trade Designations						
			Shell "Epon"	Dow "DER"	Ciba "Araldite"	Celanese "EPI-REZ"	Reichhold "Epotuf"	Gen. Mills "Gen. Epoxy"	U. Carbide* "Bakelite"
0	170–178	4–6,000 cps	825	332	6004	508	–	–	–
0.07	180–190	7–10,000 cps	826	330	6005	509	37–139	185	2772
0.14	190–200	10–16,000 cps	828	331	6010	510	37–140	190	2774
2.3	450–550	65–80°C	1001	661	7065	520	37–301	525	–
4.8	850–1000	95–105°C	1004	664	6084	530	37–304	925	–
9.4	1500–2500	115–130°C	1007	667	6097	540	37–307	1800	–
11.5	1800–4000	140–155°C	1009	–	6099	550	–	–	–
30	4000–6000	115–165°C	1010	–	–	–	–	–	–

*No longer available commercially; for reference only.

Table 3.3-11. *Aliphatic epoxy resins.*

veloped by Ciba-Geigy [33], offer low viscosity (2,500 cps), water-solubility (uncured), and higher epoxy functionality than DGEBA resins. They are useful alone or blended with DGEBA [34] and are best cured with anhydrides or aromatic amines.

Glycidyl Amines. One way to increase epoxy functionality is to base the resin on an aromatic amine instead of a phenol, thereby doubling the number of active hydrogens which may be replaced with a glycidyl group; molecules with both phenolic and amino functionality may also be used. Two of the most widely used of these specialty resins are shown below:

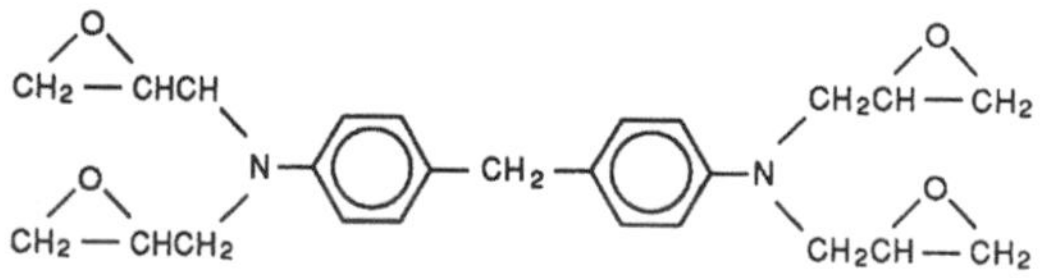

N,N,N′,N′-tetraglycidyl-4,4′-diaminodiphenyl methane WPE = 125; viscosity 10,000–15,000 cps at 50°C; commercially available as Ciba-Geigy MY-720:

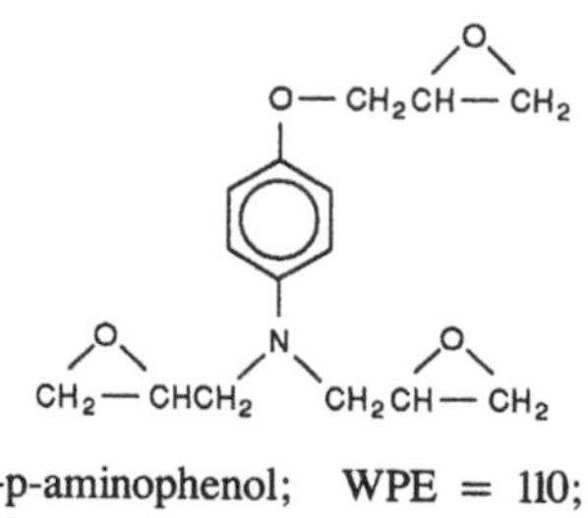

triglycidyl-p-aminophenol; WPE = 110; viscosity: 1,500–5,000 CPS; commercially available from Ciba-Geigy as Araldite 0510. These resins offer very high functionality leading to high cross-link density in the cured matrix and extremely high HDT values (to 250°C). MY-720 is the base resin for several state-of-the-art structural prepreg resin systems used in aircraft laminates (including Narmco 5208, Hercules 3501-6, and Fiberite 934). Triglycidyl-p-aminophenol is also used in combination with other resins to improve elevated temperature performance of the prepreg system.

Curing Epoxy Resins

General Characteristics of Major Curing Agents. Although the epoxide group reacts with a wide variety of groups, only three categories are widely used in curing epoxy matrix resins. These are amines, anhydrides, and catalytic curing agents.

Amines are the largest category of curing agents and include both aliphatic and aromatic amines. Aliphatic amines are relatively strong bases and are so much more reactive than aromatic amines toward epoxies that they give room temperature cures. Their short pot life makes them useful for contact molding operations but not for prepregging or filament winding. Because of their reactivity, aliphatic amines also give high exotherms in curing. Since many commonly used aliphatic amines (e.g., DETA, TETA) are relatively volatile, high exotherms accentuate toxicity and dermatitis problems by volatilizing unreacted amine. This toxicity has led to the use of various amine adducts (with epoxides, cyanoacrylates, ethylene oxide, etc.) which reduce volatility, decrease exotherms, increase amine equivalent weight, and give cured resin properties similar to those of the unmodified amines [35]. Structures of some common aliphatic amine curing agents are shown in Table 3.3-12.

Aromatic amines require elevated temperature for complete cures and are widely used in filament winding and prepreg application. The most common types are shown in Table 3.3-13. Since they are solids, the aromatic amines are more difficult to handle in mixing operations with the resin than are the liquid curing agents; typically, the resin must be heated to dissolve the solid amine. Liquid aromatic amine curing agents can be obtained from eutectic mixtures (e.g., MPDA + MDA), and they provide improved handling at no sacrifice in cured resin properties.

Cyclic *anhydrides* are useful curing agents for all epoxy resin types. They require high temperature (150–200°C) for long times (8–16 hours) to achieve optimum cures but give long pot lives and low exotherms. Resins cured with anhydrides have good high temperature stability, good chemical resistance (except to caustic), and better electrical properties than amine-cured resins. Both prepreg and filament winding applications use anhydride cures. Typical anhydride curing agents are shown in Table 3.3-14.

Catalytic agents are curing agents used at less than stoichiometric amounts (based on the epoxy group) that promote epoxy homopolymerization to achieve full cure. They are usually Lewis acids or bases and typically give long pot lives. Boron trifluoride blocked with ethyl amine (BF_3:MEA) (MEA = mono ethyl amine) is the most common Lewis acid type and is used alone or as an accelerator for aromatic amine cures. Used alone, at 1–5 phr, it requires temperatures of 150–

Table 3.3-12. Aliphatic amine curing agents.

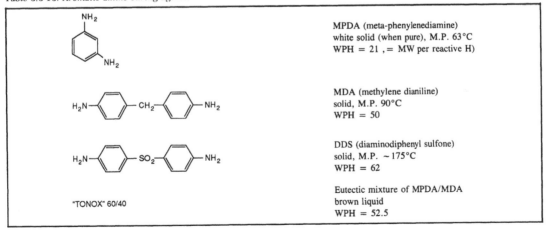

H₂NCH₂CH₂H — CH₂CH₂NH₂	Diethylene Triamine (DETA) MW 103; WPH = 21 Liquid Viscosity 6–8 cps
H₂NCH₂CH₂NCH₂CH₂N — CH₂CH₂NH₂ (with H on each N)	Triethylene Tetramine (TETA) MW 146; WPH = 24 Liquid Viscosity 20 cps
CH₃CH₂ \ NCH₂CH₂CH₂NH₂ / CH₃CH₂	Diethylaminopropyl Amine (DEAPA) MW 130; WPH = 65 Used at 4–8 phr because of catalytic activity of tert. amine
H — N(piperazine)N — CH₂CH₂NH₂	Aminoethyl Piperazine (AEP) MW 128; WPH = 43 Liquid; BP 222°C
(CH₂)ₙCNHCH₂CH₂NHR ...	Polyamidoamines High viscosity liquids or low-melting solids Used at 25–100 phr

Table 3.3-13. Aromatic amine curing agents.

(benzene ring with NH₂ and NH₂)	MPDA (meta-phenylenediamine) white solid (when pure), M.P. 63°C WPH = 21 , = MW per reactive H)
H₂N—⟨ring⟩—CH₂—⟨ring⟩—NH₂	MDA (methylene dianiline) solid, M.P. 90°C WPH = 50
H₂N—⟨ring⟩—SO₂—⟨ring⟩—NH₂	DDS (diaminodiphenyl sulfone) solid, M.P. ~175°C WPH = 62
"TONOX" 60/40	Eutectic mixture of MPDA/MDA brown liquid WPH = 52.5

Table 3.3-14. Anhydride curing agents.

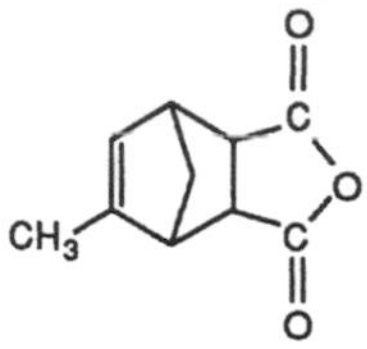

NMA: Nadic methyl anhydride
MW: 178
viscosity: 175–275 cps/25°C

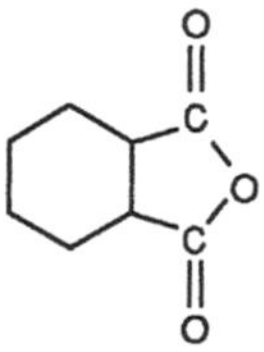

HHPA: hexahydrophthalic anhydride
MW: 154
M.P.: 35–37°C

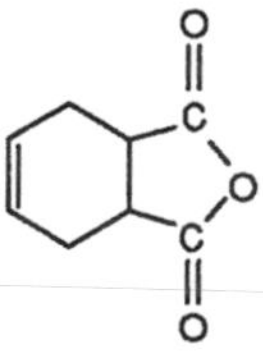

THPA: tetrahydrophthalic anhydride
MW: 152
M.P.: 99–101°C

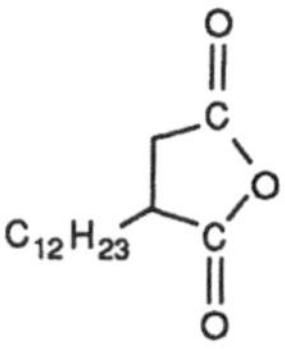

DDSA: dodecenyl succinic
MW: 266
viscosity: 290 cps/25°C

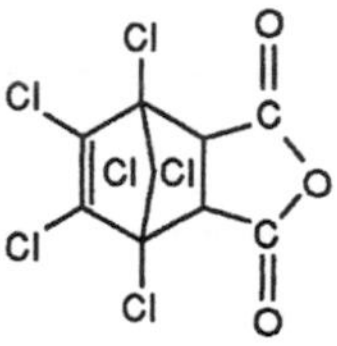

HET: chlorendic anhydride
MW: 371
M.P.: 239–240°C

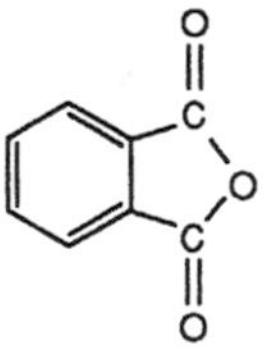

PA: phthalic anhydride
MW: 148
M.P.: 131°C

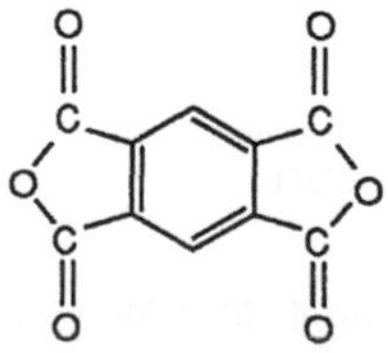

PMDA: pyromellitic dianhydride
MW: 218
M.P.: 286°C

Table 3.3-15. Catalytical and miscellaneous curing agents.

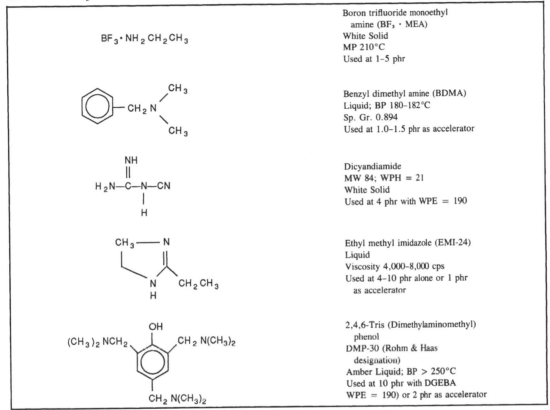

$BF_3 \cdot NH_2 CH_2 CH_3$	Boron trifluoride monoethyl amine ($BF_3 \cdot$ MEA) White Solid MP 210°C Used at 1–5 phr
	Benzyl dimethyl amine (BDMA) Liquid; BP 180–182°C Sp. Gr. 0.894 Used at 1.0–1.5 phr as accelerator
	Dicyandiamide MW 84; WPH = 21 White Solid Used at 4 phr with WPE = 190
	Ethyl methyl imidazole (EMI-24) Liquid Viscosity 4,000–8,000 cps Used at 4–10 phr alone or 1 phr as accelerator
	2,4,6-Tris (Dimethylaminomethyl) phenol DMP-30 (Rohm & Haas designation) Amber Liquid; BP > 250°C Used at 10 phr with DGEBA WPE = 190) or 2 phr as accelerator

200°C for complete cure. Thermal decomposition of the boron trifluoride-amine complex is apparently not necessary for initiation of cure [36,37]. Unlike most curing agents, $BT_3 \cdot$ MEA is a latent catalyst at room temperature and provides pot lives of several months.

Lewis bases, sometimes used alone, but most often used as accelerators for anhydride cures, are typically tertiary amines. Their effectiveness depends on structure: more steric hindrance as the nitrogen reduces reactivity and higher base strength increases reactivity [38]. Representative structures are shown in Table 3.3-15.

Cure Chemistry. Conversion of the resin/hardener mixture of low molecular weight species to a highly cross-linked network is called "cure." The objective in the curing reaction is to convert as many epoxide groups as possible. In contrast to polyesters, epoxies cure by ionic reactions, usually polyadditions, between the epoxide and the curing agent; in "catalytic" cures, the major reaction is epoxy to epoxy, i.e., homopolymerization of epoxide groups. The importance of the curing agent in epoxy reactions cannot be overstated. For a given resin, the choice of curing agent

- determines the type of curing reaction that occurs
- exerts a major influence on the time and temperature of the cure cycle
- affects final properties of the cured resin (it may account for 50% of the weight of the cured "resin")

Although there are many combinations of resin and curing agents used in composites, four basic reactions, shown in Figure 3.3-18, account for most of the cures and will be the only ones discussed here. The literature on cure of epoxies is enormous—more detailed information is available in references [35,36,39].

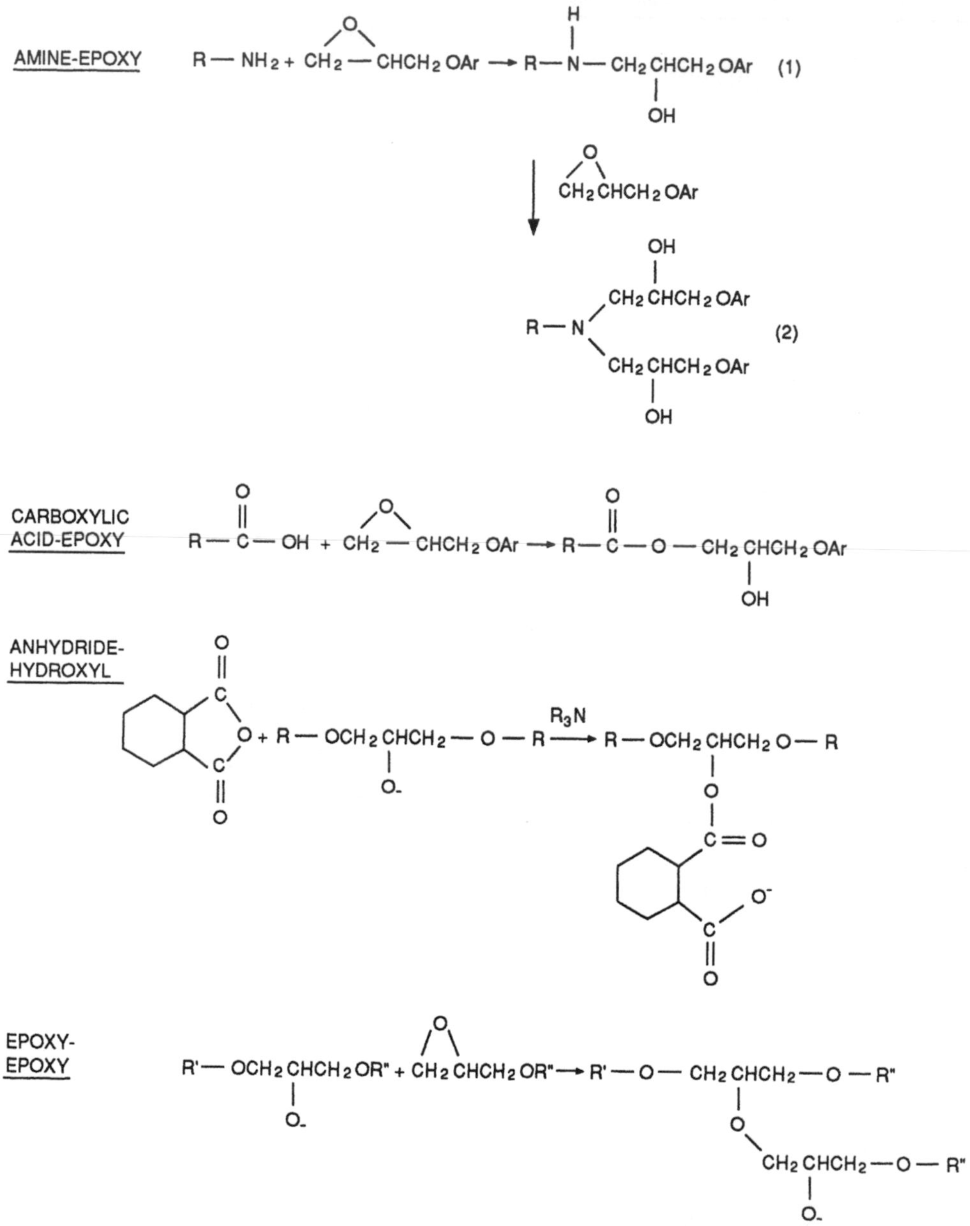

AMINE-EPOXY

$R-NH_2 + CH_2-CHCH_2 OAr \rightarrow R-N-CH_2 CHCH_2 OAr$ (1)

with H on N, OH on CH

$CH_2 CHCH_2 OAr$ (epoxide)

$R-N$ bonded to two $CH_2 CHCH_2 OAr$ groups each with OH (2)

CARBOXYLIC ACID-EPOXY

$R-C-OH + CH_2-CHCH_2 OAr \rightarrow R-C-O-CH_2 CHCH_2 OAr$
with OH

ANHYDRIDE-HYDROXYL

anhydride $+ R-OCH_2 CHCH_2-O-R \xrightarrow{R_3N} R-OCH_2 CHCH_2 O-R$
with O^-
product with $C=O$, O^- and cyclohexane ring, $C=O$

EPOXY-EPOXY

$R'-OCH_2 CHCH_2 OR'' + CH_2 CHCH_2 OR'' \rightarrow R'-O-CH_2 CHCH_2-O-R''$
with O^-
O
$CH_2 CHCH_2-O-R''$
O^-

FIGURE 3.3-18. Basic curing reactions.

Amine-Epoxy. The only significant reaction occurring between aliphatic or aromatic amines and the epoxy group [40–42] is the amine hydrogen-epoxide ring reaction shown in Figure 3.3-18. For aliphatic amines, addition of the first amine hydrogen is followed closely by addition of the second [40] hydrogen without heating so that room temperature cure occurs. If the amine is aromatic (or if the aliphatic amine is sufficiently hindered), the reaction stops at Step 1 and the product is the soluble, chain-extended amino alcohol, commonly referred to as a "B-staged" resin. Further heating (to 150–180°C) is required to complete the reaction.

The cured product (structure 2 in Figure 3.3-18) contains a tertiary amine which is so sterically hindered that it does not cause further reaction of epoxide groups [40]. It also contains one hydroxyl group for each amine hydrogen that reacts; these hydroxyls act as accelerators for the amine-epoxy reaction [43] but do not participate in etherification reactions during any stage of aromatic or aliphatic amine-epoxide cure [40,41]. The five-atom repeat group formed from the amine-epoxy reaction,

$$(An-O-CH_2\,CHCH_2-N-Ar)$$
$$|$$
$$OH$$

is characteristic only of amine-cured epoxies and accounts for some observed differences between amine cures and other types; examples include the γ-relaxation process attributed by some [44,45] to the $(-O-C-C-C-)$ portion and thermal degradation of
$$|$$
$$OH$$
amine-cured epoxies [46,47].

Carboxylic Acid-Epoxy. Carboxylic acids react with the epoxy group by base-catalyzed addition esterification, as shown in Figure 3.3-18, to form the hydroxy ester [48]. Although carboxylic acids are not typically used alone to cure epoxies, the reaction is part of the cure sequence in anhydride-epoxy systems. In addition, the same addition esterification reaction is used to manufacture vinyl ester resins [14].

Acid Anhydride-Hydroxyl. Carboxylic acid anhydrides, widely used in curing epoxy resins for reasons already discussed, react with epoxies by a combination of anhydride-hydroxyl reactions to give the half acid ester, which then reacts with an epoxide ring by the addition esterification [14–16] reaction previously mentioned. The overall sequence is illustrated in Figure 3.3-19; the repeating unit formed in this sequence is a polyester of a substituted ethylene glycol and the acid anhydride used. The actual cure sequence is complicated by a competing etherification reaction which is minimized by tertiary amine catalysis [17] and by higher cure temperatures [18]. Anhydride cures are also affected more by moisture in the system than are aromatic amine cures, as might be expected if water attacks the anhydride. Moisture may be introduced into the system not only by the resin itself but also by the reinforcement or filler. The moisture absorbed in undried aramid fiber reinforcements, for example, may lead to complete disruption of anhydride cures.

Epoxy-Epoxy. In contrast to the other three basic reactions, which are all polyadditions of a "curing agent" to the epoxy group, the epoxy-epoxy reaction is a homopolymerization of the epoxy resin. It may be catalyzed by Lewis acids or bases, but both give the same polyether structure [17] shown in Figure 3.3-18. The most important commercial example of catalytic cures by Lewis acids is the use of $BF_3 \cdot MEA$. Imidazoles are the most important examples of Lewis bases used alone for catalytic cures. Some epoxy-epoxy etherification occurs in reactions with other curing agents such as diethylaminopropyl amine [42] and presumably in cures with dicyandiamide (used at less than stoichiometric for amine hydrogen/epoxy).

Cured Resin Properties

In spite of the enormous number of resin and curing agent combinations in use, a few generalizations can be made about "epoxy" neat resin properties:

- Multifunctional resins (novolacs, tetrafunctional glycidylamines, etc.) are necessary to achieve high HDT (>200°C).
- Standard liquid DGEBA resins cured with typical aromatic diamines or anhydrides give HDT values up to about 175°C; aliphatic amines give HDT values up to 110–120°C (see Tables 3.3-16 and 3.3-17).
- High temperature, long cures are usually necessary to get the highest HDT values.
- The presence of diluents lowers HDT without major effects on room temperature mechanical properties (Table 3.3-17).

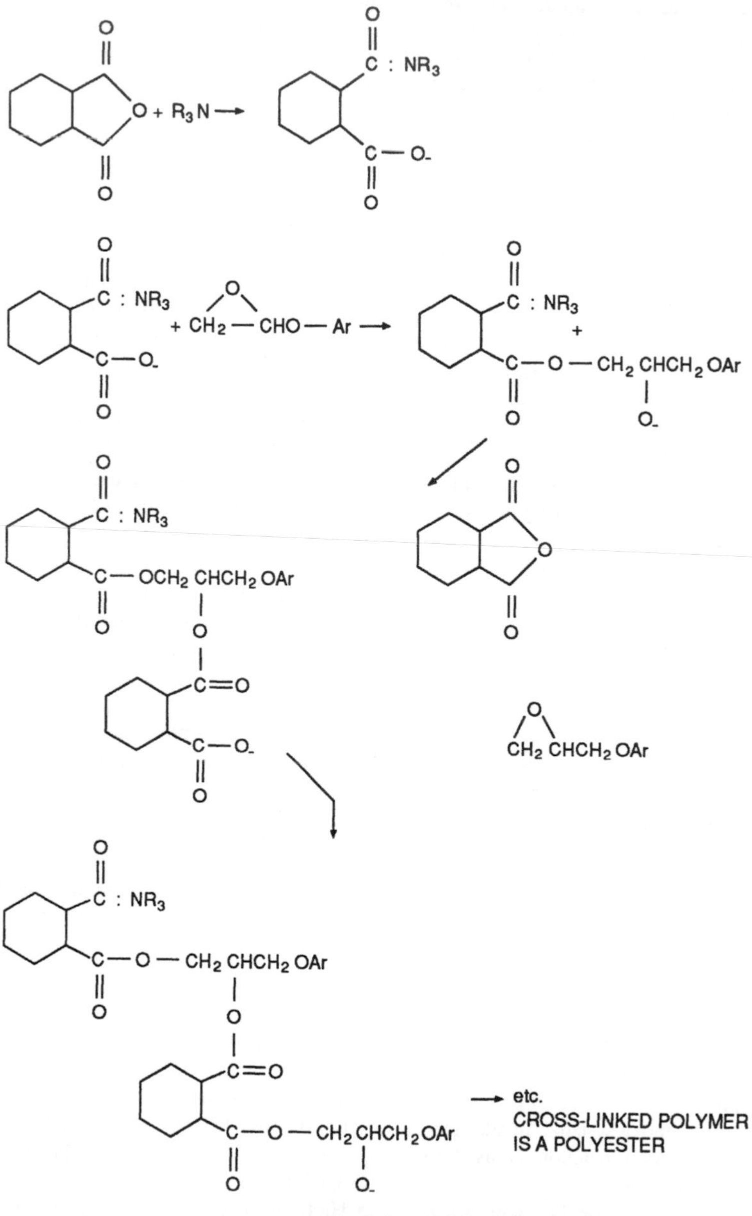

FIGURE 3.3-19. Anhydride curing reactions.

Table 3.3-16. Cured resin properties; mechanical, chemical, thermal, and electrical performance of a standard DGEBA epoxy (WPE = 190) cured with anhydride, aromatic amine, and Lewis acid curing agents.

Cure Cycle		NMA 2 hr/90° +4 hr/165° +16 hr/200°	MDA 16 hr/55° 2 hr/125° 2 hr/175°	BF₃ 4 hr/100° +16 hr/150°
HDT	°C	156	160	168
Flex strength	10³ psi	14.0	13.5	14.5
Flex modulus	10³ psi	440	390	450
Comp. yield str.	10³ psi	18.3	16.8	16.5
Comp. modulus	10³ psi	440	375	330
Tensile str.	10³ psi	10.0	10.2	5.7
Ult. elongation,	%	2.5	4.4	1.6
% wt. loss, 500 hr/210°C		1.8	6.1	5.5
% wt. change, 120 days rt				
in: 10% NaOH		+0.5	−0.05*	+1.46
30% H₂SO₄		+0.55	+2.7	+1.20
toluene		+0.28	+0.25	+0.26
distilled H₂O		+0.87	+2.0	+1.8
Dielectric Constant				
60 Hz		3.15	4.10	3.47
10³		3.14	4.06	3.45
10⁶		2.97	3.56	3.23
Dissipation Factor				
60 Hz		0.0020	0.0054	0.0029
10³		0.0054	0.015	0.0053
10⁶		0.017	0.036	0.023
C.A. Conc., PHR		87.5	26	3

*In 50% NaOH.

Table 3.3-17. Comparison of diluted and undiluted DGEBA resin cured by aliphatic and aromatic amine.

	Curing Agent			
	TETA		MPDA	
	DGEBA (DER331)	DGEBA + diluent (DER334)	DGEBA (Epon 828)	DGEBA + diluent (Epon 815)
Curing Agent, phr	13	13.5	14.6	14.5
Mix Viscosity	2250	280	–	–
Cure Cycle	16 hr/25°C +3 hr/100°C		2–6 hr/80°C +6 hr/160°C	
HDT, °C (264 psi)	111	84	150	105
Tensile				
Ult. Strength, ksi	11.4	10.7	12.9	12.4
Ult. Elongation, %	4.4	3.8	6.5	10.6
Modulus, ksi	–	–	462	430
Compressive				
Yield Strength, ksi	16.3	14.3	18.6	17.4
Modulus, ksi	440	305	–	–
Flexural				
Strength, ksi	13.9	11.8	20.5	20.6
Modulus, ksi	440	404	470	480

Table 3.3-18. Thermal stability of epoxy resins cured
with aromatic amines and imidazoles [50].

DGEBA Resin, parts (Epi-Rez 510)	100	100
Curing Agent, phr		
Imadazole (EMI-24)	2	–
Aromatic Amine (MPDA)	–	14.2
Cure Cycle	2 hr/93°C +2 hr/150°C +2 hr/200°C	
% Weight Loss at 200°C (in air)		
1 day	1.2	1.5
4 days	2.1	2.5
7 days	2.5	3.2
% Weight Loss at 260°C (in air)		
1 day	2.0	9.1
4 days	5.1	decomposed

- The use of flexibilizing units (in the resin or in the curing agent) gives higher elongation and improved impact strength at the expense of lower HDT, tensile strength and modulus, and compressive yield strength.
- The choice of a curing agent for a given resin is

likely to have a greater effect on chemical resistance and on electrical and thermal properties than on mechanical properties. The use of Nadic methyl anhydride instead of m-phenylene diamine in curing a liquid DGEBA resin, for example, results in a drastic reduction in base resistance and significantly improved dielectric and thermal aging properties at very little change in strength and modulus of the resin (see Table 3.3-16).

The higher service temperatures for anhydride vs. aromatic amine cures [49] are presumably due to the lower thermal stability of the amino-alcohol cure link formed by the amine-epoxy reaction [46]. A comparison of the weight loss data for imidazole- vs. aromatic-cured resins in Table 3.3-18 suggests that the polyether repeat formed from the imidazole-catalyzed epoxy homopolymerization is also more thermally stable than the amine cure link. Data in reference [50] also show that resistance of the imidazole-cured epoxy to acid, base, and solvent is at least as good as that of the aromatic amine-cured epoxy.

Cured resin properties for three "wet" filament wind-

Table 3.3-19. Properties of epoxy filament winding systems.

Formulation	1	2	3
Resin (100 parts)	826	CY-179	DER383
Diluent (phr)	RD-2 (25)	–	–
Other (phr)	–	Hycar 1300 × 8 Rubber Modifier	–
Curing Agent (phr)	Tonox 60/40 (30)	MTHPA (100)	MTHPA (80)
Catalyst (phr)	–	1MI (1)	DMP-30 (3)
Mix Viscosity, cps	1,000–1,200	380	740° (at 34°C)
Cure Cycle	3 hr/60°C +2 hr/120°C	2 hr/90°C +4 hr/150°C	2 hr/90°C +4 hr/150°C
HDT, °C (264 psi)	121	170	127
Tensile			
Ult. Strength, ksi	13	8.4	11.5
Ult. Elongation, %	9	2.8	4.0
Modulus, ksi	389	348	–
Compressive			
Strength, ksi	16.0	17.4	–
Strain, %	8.0	8.4	–
Modulus, ksi	420	368	–
Flexural			
Strength, ksi	–	–	21.7
Modulus, ksi	–	–	470
Reference:	[51]	[52]	[53]

ing systems are shown in Table 3.3-19. "Wet" winding means that the fiber bundle is impregnated with liquid resin just prior to winding on the mandrel, rather than being prepregged on ahead of time. A major problem in formulating a resin system for this process is achieving a viscosity low enough for good fiber bundle wet-out, while maintaining both the balance of properties needed (HDT, strength, resiliency) in the cured resin and the necessary pot life to complete the process. The examples in Table 3.3-19 illustrate the following:

- The use of a fairly low viscosity resin with just enough diluent to reduce the viscosity to an acceptable level and an aromatic amine curing agent. In spite of its difunctionality, the diluent causes a significant decrease in HDT (Formulation 1).
- The use of a more highly functional base resin plus rubber modifier to improve elongation and toughness and an anhydride curing agent which helps reduce mix viscosity because of the large amount of low viscosity anhydride needed for stoichiometry. HDT remains high at the expense of elongation (Formulation 2).
- The usc of a low viscosity DGEBA resin alone with only the anhydride to further reduce viscosity. HDT, elongation, and viscosities are intermediate between the other two (Formulation 3).

Further results on cure and viscosity characteristics of epoxies are provided in references [54–56].

Moisture Absorption

Moisture absorption of epoxy resins most widely used in composites is usually in the 1.5–6.0% range. As with most polymers, the amount of water absorbed depends on the relative humidity, as shown in Figure 3.3-20, for two state-of-the-art structural epoxy formulations widely used in prepregs. The rate at which moisture is absorbed depends on the diffusion coefficient (D) for the resin in question, and D, in turn, is temperature dependent. If the absorption is governed by Fick's law, a plot of % weight gain vs. (time)$^{1/2}$ gives the curve shown in Figure 3.3-21, consisting of an initial straight line portion followed by a leveling off at the equilibrium absorption value. Plots of this type can be used to determine the diffusion coefficient for the resin (or composite) at a given relative humidity and temperature. Diffusion coefficients in the order of 10^{-7}

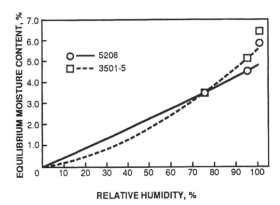

FIGURE 3.3-20. Equilibrium water absorption for epoxy resins [57].

mm²/sec have been reported for typical structural grade epoxy resins at 30–70°C [57–59].

3.3.4 Thermoplastic Matrix Resins

Introduction

Although thermoplastic resins account for over 85% of all plastic resin sales, their use in reinforced plastics is relatively low, amounting to about 15–16% of all matrix resins. Most use of thermoplastic resins is in short-fiber reinforced thermoplastic resins processed by

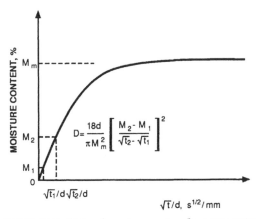

FIGURE 3.3-21. Water absorption vs. time for epoxy resins [57].

Table 3.3-20. Properties of glass reinforced engineering thermoplastics (short-fiber, injection molded).

	% Fiber (wt)	Tensile Strength 10^3 psi	Tensile Elongation %	Flexural Strength 10^3 psi	Flexural Modulus 10^6 psi	Izod Impact Strength ft-lb/in Notched	HDT, °C 264 psi
Polysulfone Udel®	30	18.0	3	24.0	1.20	1.8	185
Polyethersulfone Victrex® PES 430P	30	20.3	3	27.5	1.22	1.4	216
Polyetherether ketone Victrex® PEEK	20	19.6	4.4	–	1.52	–	286
Polyetherimide Ultem® 2300	30	24.5	3	33.0	1.20	2.0	210
Poly(amide-imide) Torlon® 5030	30	28.3	5	46.1	1.61	2.0	274
Polyphenylene Sulfide Ryton® R-4	40	19.5	1.3	29.0	1.70	1.4	>260
Polybutylene Terephthalate	30	20.0	4	28.0	1.20	2.6	221

Data from suppliers' literature.

injection molding and extrusion. In these applications, the reinforcement is present primarily to upgrade the properties of the matrix resin (usually its modulus), and reinforcement levels are relatively low (10–30% by vol.). Almost all thermoplastic resins are now available commercially in ready-to-mold reinforced versions. Properties of short-fiber, injection molded thermoplastic matrix composites are given in Table 3.3-20.

In spite of the wide use of thermoplastics with short fibers, there is currently a significant emphasis on applications in which the thermoplastic resin is used as a matrix resin for a continuous fiber, high-performance composite (e.g., replacing an epoxy resin) [60]. The purpose of this subsection is to examine thermoplastic resins which appear to be suitable for such use and to review some of the published work in which thermoplastic resins have been evaluated as matrices for continuous fiber composites.

Characteristics of Thermoplastic vs. Thermosetting Resins

Thermoplastic resins are collections of high molecular weight (> 15,000) linear or branched molecules which flow upon the application of heat and pressure and are usually soluble in a suitable solvent. Since they can be melted and reformed into another shape, thermoplastics are processed almost exclusively in the melt.

Thermosetting resins are formed from relatively low molecular weight precursor molecules which are reacted ("cured") to form a cross-linked polymer network that does not flow under heat and pressure and is not soluble. Once cured, the thermosetting polymer is therefore intractable. Thermosetting matrix resins are available in a wide range of formulations having T_g values (glass transition temperature) from 45–300°C and elongations of 1% to > 100%. Typical difunctional epoxy resins (DGEBA type) cured with aromatic amines have T_g values in the 130–160° range and ultimate elongations of 4–8%; to increase the T_g to the 160–225°C range, it is necessary to use combinations of DGEBA plus multifunctional resins (e.g., epoxy novolacs) or, for the upper portion of this range, to use exclusively multifunctional resins combined with highly stable curing agents. An example of the latter type is the current state-of-the-art structural epoxy which is supplied by several prepreggers in formulations that are all based on a tetrafunctional epoxy (tetraglycidyl methylene dianiline) plus an aromatic diamine (4,4′-diaminodiphenyl sulfone) and minor ingredients. Commercial examples include Hercules 3501-6 and Narmco 5208. Figure 3.3-23 shows stress-strain curves for 3501-6 resin, a DGEBA/MPDA resin,

Generic Name Trade Name Manufacturer	HDT °C (264 psi)	Structure
Polycarbonate Lexane®, Merlon® GE, Mobay	133	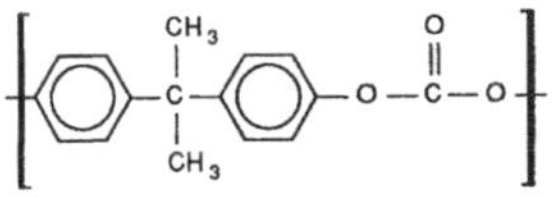
Polyphenylene Sulfide Ryton® Phillips	137	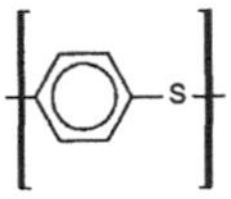
Polyetherether Ketone Victrex® PEEK ICI	148	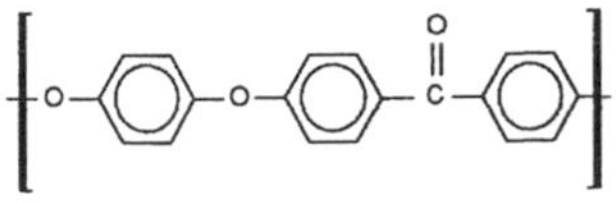
Polysulfone Udel® Union Carbide	174	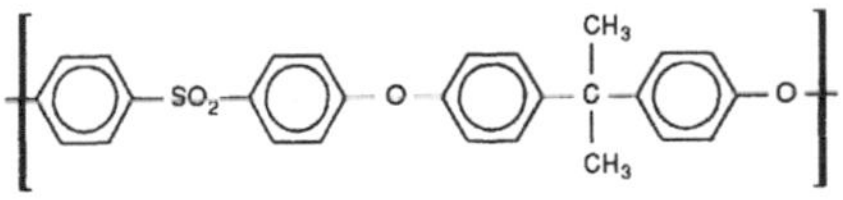
Polyarylate Ardel® Union Carbide	174	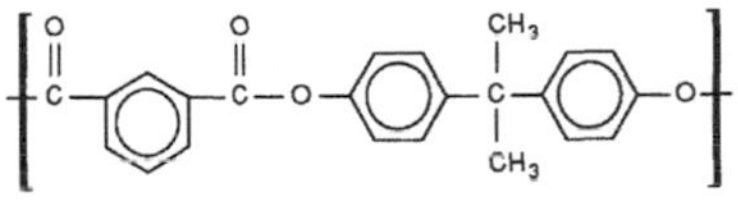
Polyetherimide Ultem® GE	204	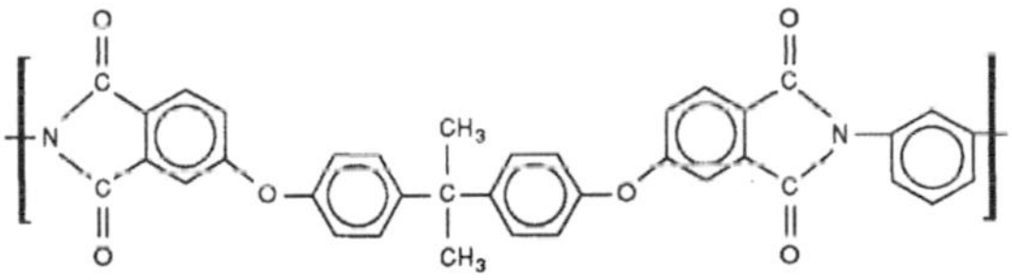
Polyphenylsulfone Radel® Union Carbide	204	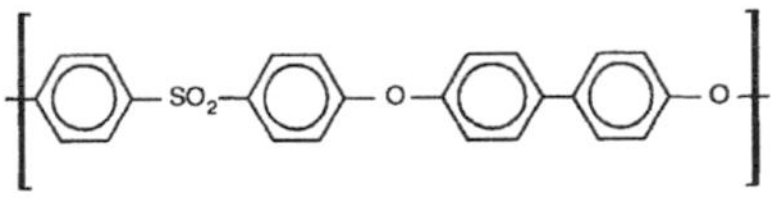
Polyethersulfone Victrex® PES ICI	204	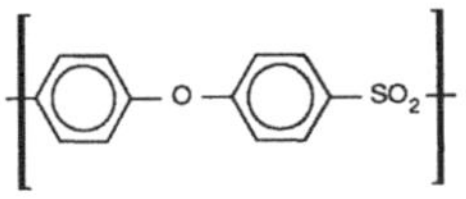
Polyamide-imide Torlon® Amoco	274	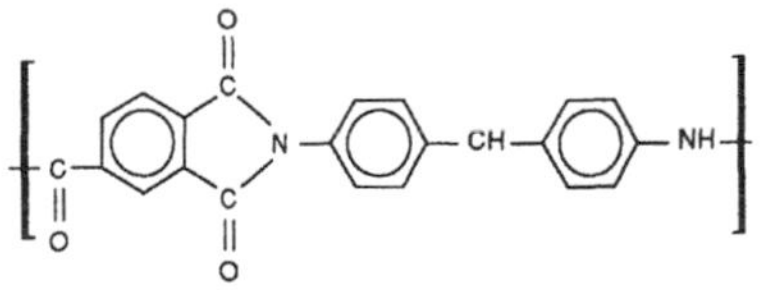
Polyarylsulfone Astrel® 3M	274	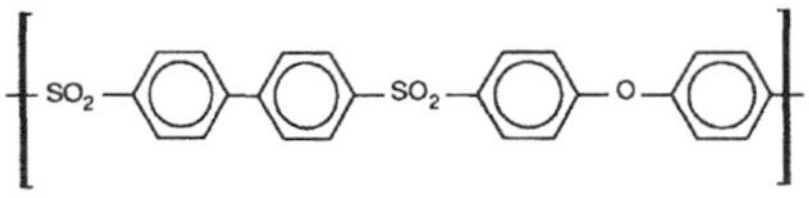

FIGURE 3.3-22. High temperature engineering thermoplastics.

and two thermoplastic resins. The curves illustrate one of the major limitations of this approach to achieving a high T_g resin, namely, low strain-to-failure (about 1.5%) and generally brittle behavior which manifests itself in poor impact strength and low damage tolerance (e.g., little ability to carry compressive loads after impact damage). To make matters worse, the particular formulations in use are susceptible to plasticization by absorbed moisture, which lowers the resin T_g, thereby partially negating the major benefit that the high cross-link density contributed. These factors, along with the long cure cycles, have contributed to the interest in alternative matrix resins [60].

Thermoplastic resins, by contrast, achieve high T_g by virtue of stiff, linear chains and high molecular weight, and are generally stronger, tougher, and more resistant to moisture than the highly cross-linked epoxies.

Why Use Thermoplastics?

There is an extremely high potential for application of thermoplastic resins in the near future. The reasons for this interest fall into two broad categories:

- cost savings from improved processing and handling compared to epoxies

- property improvements stemming mainly from increased toughness and lower moisture sensitivity vs. the epoxies

The savings in the first category are expected to come from both materials and processing costs vs. the incumbent epoxy systems. Overall manufacturing costs for the thermoplastics should be lower than for the epoxy incumbents due to a number of factors:

- adaptability of thermoplastics to high rate processing, thermoforming being inherently faster than typical cure cycles
- simpler storage and handling of materials, indefinite shelf life without refrigeration, simpler incoming quality control
- reduced scrappage rate because of reformability and the potential for using scrap in other operations (e.g., molding compounds)
- fewer repairs because of the higher toughness and impact resistance of thermoplastics, which are generally more ductile

In an Air Force contract study of fabrication costs for several designs (body section, wing box, fuselage panel, etc.), Boeing compared graphite/epoxy with

Table 3.3-21. Polysulfone vs. epoxy fabrication costs [61,62].

Material	No. of Units	Production Hours	Material $	Tooling Hours	Total Cost Relative	Total Cost $
		Example 1: Aircraft Fuselage Panel **Example 2: Compass Cope Horizontal Stabilizer**				
Example 1						
Graphite/	1	1,425	2,452	1,310	1	84,502
Epoxy	10	10,140	24,520	1,310	4.57	386,020
($25/lb)	100	62,349	245,200	1,310	25.5	2,154,970
Graphite/	1	961	2,099	2,184	1.14	96,449
Polysulfone	10	6,838	20,990	2,184	3.45	291,650
($25/lb)	100	42,048	209,900	2,184	18.2	1,536,860
Example 2						
Glass/	1	1,068	600	810	1	56,940
Epoxy	10	7,600	6,000	810	4.54	258,296
($12/lb)	100	46,730	60,000	810	26.1	1,486,178
Graphite/	1	775	2,760	1,690	1.35	76,710
Polysulfone	10	5,515	27,600	1,690	4.28	243,747
($65/lb)	100	33,909	276,000	1,690	23.6	1,343,980

graphite/polysulfone and found significant cost savings for the latter beginning at only a few units and increasing with volume up to the 100-unit level. In every case, lower production time accounted for most of the savings for the polysulfone system. Although both the graphite/epoxy and graphite/polysulfone were assigned the same unit costs, the overall material costs for the thermoplastic system were slightly lower for each level of every case because of associated savings assumed for it (e.g., scrappage, storage losses). On the other hand, tooling costs were considered to be about 70% higher for the thermoplastic system. Examples of this analysis for a fuselage panel and a stabilizer are shown in Table 3.3-21.

Property improvements in parts are expected to result from the inherently tougher thermoplastic resin compared to its epoxy counterpart [63]. In addition to better end use performance, higher toughness should lead to fewer repairs. The higher strain-to-failure of thermoplastics (vs. incumbent epoxies) is another desirable manifestation of their overall toughness advantage (Figure 3.3-23). Most high-performance thermoplastics are less affected by moisture than are the state-of-the-art, high temperature epoxies; this moisture tolerance is expected to result in lower knock-down factors for hot/wet conditions.

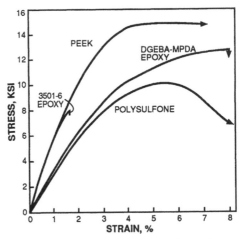

FIGURE 3.3-23. Tensile stress vs. strain for thermoplastic and epoxy resins.

Some identified and potential disadvantages of thermoplastics which must also be kept in mind include

- solvent sensitivity—Most high temperature thermoplastics now available are amorphous and are more solvent-sensitive than epoxies. If a semicrystalline polymer (e.g., PPS, PEEK) with good solvent resistance is chosen, preimpregnation of the reinforcement from solution may also be eliminated.
- boardy prepregs which are difficult or impossible to handle in some established processes
- considerably higher processing temperatures required to achieve a given use temperature than is usual for epoxies
- higher resin (and prepreg) costs—Currently available thermoplastics which cost less than baseline structural epoxy systems do not meet all property requirements. This relation could, of course, change, especially if sufficient volume were to develop for the selected thermoplastic by its being used in other applications.

Comparison of Currently Available Thermoplastics

Structure Comparisons. The term *engineering thermoplastics* is used loosely to categorize a group of resins which have generally better mechanical properties and high temperature capability than the "com-

Table 3.3-22. Engineering thermoplastics, annual USA consumption by resin type [64].

	1981 Volume 10⁶ lb	Cost (10–82)	
		¢/lb	¢/in³
Nylon	275	181	7.4
Polycarbonate	250	166	7.2
Mod. PPO	125	139–237	5.4–9.0
Acetal	95	148	7.6
TP polyester	55	136–170	6.4–8.0
Polysulfone	10	382	17.1
Polyphenylene sulfide	7	320 (40% glass)	18.5
Others	1		
Total	818		

Notes:
1. Worldwide use of engineering TP is about 1.9 × 10⁹ lb.
2. Approx. distribution: U.S.   42%
                          Europe   40%
                          Japan   18%
                                  100%
3. Annual growth rate ~7–8%.

modity" resins (polyethylenes, polypropylene, poly-styrene, polyvinyl chloride, etc.). Structurally, the commodity resins are mostly addition polymers having only carbon–carbon bonds in the backbone repeat unit, whereas the engineering resins all have heteroatoms and/or one or more rings in the repeat unit. The most important engineering resins are listed in Table 3.3-22 in order of USA sales volume in 1981. Total worldwide consumption of this group of resins is about 2 billion pounds, or only 5–6% of the total plastics resins sold.

The structures of the high temperature resins which are matrix candidates for composites are shown in Figure 3.3-22, arranged in order of increasing heat deflection temperature. One of the resins, Astrel 360, is no longer commercially available but is included here because it has received a fairly complete evaluation as a matrix resin for graphite in several studies. With two exceptions, all resins shown in the figure are amorphous; the exceptions, polyphenylene sulfide (PPS) and polyetherether ketone (PEEK), are semicrystalline polymers. A common structural feature in all the resins is the presence of a high proportion of aromatic rings, usually connected by a stable heteroatom or group $(-SO_2-, -O-, -\overset{\|}{\underset{O}{C}}-$, etc.); this group leads to a high degree of chain rigidity, which contributes to the high glass transition temperature of these resins. In addition, all the resins have the low aliphatic hydrogen (C–H) content which is necessary for good thermal stability at high temperatures.

Mechanical Properties. Table 3.3-23 summarizes properties for the high temperature thermoplastics. Tensile strengths are generally in the 10–15 ksi range and tensile moduli in the 360–500 ksi range. The corresponding values for a state-of-the-art structural epoxy (Hercules 3501-6) are $TS = 8$ ksi and $TM = 600$ ksi. For a typical diglycidyl ether of bisphenol A/aromatic amine epoxy (Epon 828/MPDA), the corresponding values are $TS = 13$ ksi and $TM = 400$ ksi. Stress-strain curves for polysulfone, PEEK, and these two epoxies are compared in Figure 3.3-23. The low failure strain of the 3501-6 epoxy is a result of its high cross-link density, which is necessary to achieve a high glass transition temperature. Both the DGEBA epoxy and polysulfone show much higher ultimate strain capability. The polysulfone curve is typical for most of the other ductile thermoplastics listed in Table 3.3-23, having a well-defined yield point at 5–6% strain and an ultimate elongation of over 50%. PEEK has an initial modulus essentially the same as that of 3501-6 epoxy, yields at a stress almost two times higher than the failure stress for the epoxy, and fails at about 100 times the failure strain for 3501-6. The enormously larger area under the PEEK stress-strain curve represents a degree of toughness which shows up in the composite and will be discussed in a later subsection.

The ductile behavior in tension and the Izod impact strength values over 1.0 ft-lb/in (notched) are both indicative of better overall toughness for the thermoplastics in Table 3.3-23. An exception is polyphenylene sulfide, which is quite brittle in the unreinforced state

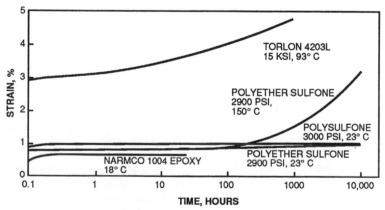

FIGURE 3.3-24. Creep of thermoplastic and epoxy neat resins (thermoplastic data from suppliers, epoxy data from reference [66]).

Table 3.3-23. *Neat resin properties.*

Property		Udel®	PES	Radel® R-5000	Astrel®	PEEK	Ryton®	Torlon®	Ultem®	Ardel®	Valox®	Lexan®
Tensile												
Yield Str.	ksi	10.2		10.4					15.2	9.5		
Ultimate Str.	ksi		12.2		13.1	14.5	10.8	26.9			8.0	9.0
Yield Elong.	%	5–6		7	13	6	–	–	7–8	8	–	7
Ult. Elong.	%	50–100	40–80	60		150	3	12–18	60	50	300	
Modulus	ksi	360	350	315	370		480	730	430	290		345
Flexural												
Strength	ksi	15.4	18.7	12.5	17.2		20.0	30.0	21.0	11.0	12.8	13.5
Modulus	ksi	390	373	331	395	551	550	660	480	310	340	340
Compressive												
Strength	ksi				17.9		16.0	40.0	20.3		13.0	12.5
Modulus	ksi	374			340			413	420		–	
Izod Impact Str. (notched, ft-lb/in)		1.3	1.6	12	3		0.2	2.5	1.0	4.2	1.2	15
Heat Deflection Temp., 264 psi, °C		175	203	204	274		137	274	200	174	54	133
UL Continuous Service, °C Temp. Rating		150	180						170			
Rockwell Hardness		M-69	M-88		M-110		R-124		M-109		R-117	
Specific Gravity		1.24	1.37	1.29	1.36	1.32	1.34	1.40	1.27	1.21	1.31	1.20
Coeff. Linear Thermal Exp., ppm/°C		56	55	55	47		54	36	62		95	
*Morphology		A	A		A	SC	SC	A	A	A	SC	A
Tg °C			225			143						
M.P. °C						334						
Cost, $/lb		3.82	4.25	90.00		28.00		14.50	4.25	2.60	1.50	1.64
Flammability & Smoke												
Oxygen Index		30	38	38		35	44	43	47	34		25
UL 94 Rating		V-O	V-O	V-O		V-O	V-O	V-O	V-O	V-O		
Thickness, in		0.25	0.063	0.030		0.063		0.120	0.060	0.062		
NBS Smoke, Flaming												
D_s at 4 min			0			10		2	0.7			
D_m at 20 min			30	35		–		169	30			130
Electrical												
Dielectric Constant												
10^3 H₃			3.5			3.4	3.1		3.15	2.71		
10^6 H₃		3.03	3.5	3.45			3.1		–	2.62	3.1	
Dissipation Factor												
10^3 H₃							0.0005		0.0013	0.005		
10^6 H₃		0.0034	0.0035	0.0076			0.0009		–	0.02	0.02	
Dielectric Strength (ASTM D-149)												
Volts/mil		425	400	325	1600		380	600		400	590	
Thickness, in		0.125		0.125	0.001			0.125		0.125	0.063	
Volume Resistivity (ohm-cm)		5 × 10^{16}	10^{17} – 10^{18}	3.5 × 10^{15}	3 × 10^{16}		4.5 × 10^{16}				4 × 10^{16}	
Arc Resistance, sec.		122		41				125	128	125	190	
Chemical												
Soluble in:		CH₂Cl₂	CH₂Cl₂			H₂SO₄	–		CH₂Cl₂	CH₂Cl₂		
Attacked by:							–					
H₂O Absorp., Equil., RT, %		0.7	2.1	1.16		0.3		0.28	1.25		0.34	
Grade (designation for unfilled resin)		P-1700	300P	R-5000	360			4203	1000	D-100	310	

*A = amorphous; SC = semicrystalline.

Table 3.3-24. Properties of injection molded neat resins [67].

	Test Temp. °F	Polyetherether ketone (PEEK)		Polyetherimide (Ultem®)		Poly (amide-imide) Torlon®	
			% Retention		% Retention		% Retention
Tensile Strength 10^3 psi	73	15.0	–	16.6	–	20.0	–
	300	6.9	46	8.2	49	13.0	65
	400	–		4.0	24	7.0	7
	450	2.4	16	–		–	–
Tensile Modulus 10^3 psi	73	6.1		490		650	
Tensile Elongation, %	73	4.4		6.5		5.5	
Flexural Strength 10^3 psi	73	19.3	–	20.3	–	21.7	–
	300	6.8	35	10.9	54	19.0	88
	400	1.6	8	2.5	12	6.0 (500°F)	28
Flexural Modulus 10^3 psi	73	550	–	510	–	610	–
	300	360	66	430	84	520	85
	400	40	7	280	55	300 (500°F)	49
Shear Strength 10^3 psi	73	13.8		14.3		15.7	
Shear Modulus 10^3 psi	73	170		180		170	
Poisson's Ratio	73	0.36		0.34		0.39	

and is offered commercially only as a reinforced molding compound. Studies of three neat sulfone polymers showed all to be much tougher than epoxies [65]. The fracture energy ranking was as follows:

Polyphenylsulfone (Radel®) > Polysulfone (Udel®) >

Polyethersulfone (Victrex® PES)

Creep resistance is expected to be better for cross-linked resins than for thermoplastics, but several of the high-performance thermoplastics have outstanding creep resistance, approaching that of the usual epoxy matrix resins [60]. Creep data for polysulfone, polyethersulfone, and Torlon® are plotted in Figure 3.3-24; total strain for polysulfone and polyethersulfone (at 3,000 psi and 2,900 psi) after 10,000 hours at room temperature is about 1%. Increased temperature and/or load accelerate the creep rate, as shown in Figure 3.3-24 for PES at 23°C and 150°C at the same load. Room temperature creep for an epoxy (350°F curing) is shown for comparison and is only slightly lower than

that of the two polysulfones. Temperature dependence of short-term properties of some thermoplastic resins is indicated in Table 3.3-24.

Chemical and Environmental Resistance. In general, thermoplastic resins have poorer resistance to chemicals than do thermosetting resins, but there are notable exceptions (e.g., various halogenated thermoplastics). The most important commercial resins for high chemical resistance are bisphenol A/fumarate, vinyl esters, isophthalic polyesters, and epoxies. In certain environments, some thermoplastics may outperform even these thermosets and are often used as liners which are overwrapped with a thermosetting matrix resin. For aircraft applications where thermoplastics are now being considered as matrix resins, the chemical resistance requirements are comparatively modest but are sufficient to cause problems for several otherwise satisfactory candidates.

The chemicals usually most troublesome are chlorinated solvents used as paint strippers and phosphate esters used as hydraulic fluids. Unfortunately, the various thermoplastic resin candidates have mostly been amorphous (glassy) polymers which are particularly susceptible to environmental stress crazing and cracking in these environments. This susceptibility is illustrated in Table 3.3-25, where polysulfone and two other amorphous, high temperature polymers (polycarbonate and modified polyphenylene oxide) are compared to two semicrystalline polymers (66 nylon and polyphenylene sulfide) and a thermoset (phenolic) in a number of chemical environments. In chlorinated solvents and in most of the highly polar organic solvents, the three amorphous polymers are very poor compared to the two semicrystalline polymers and the thermoset. The similarity in chemical resistance of the semicrystalline and cross-linked polymers is due largely to the physical inaccessibility of the crystalline regions of the polymer to most penetrants [69].

Table 3.3-25. Chemical resistance of high-performance resins [68].

	% Tensile Strength Retention after 24 hrs at 200°F					
Environment	66 Nylon Zytel® 101	Polycarbonate Lexan® 141	Polysulfone Udel®	Mod. PPO Noryl®	PPS Ryton®	Phenolic Genal® 4300
Acids						
85% H_3PO_4	0	100	100	100	100	73
Glacial acetic (100%)	0	67	91	78	98	98
Conc. HCL (37%)	0	0	100	100	100	83
30% H_2SO_4	0	100	100	100	100	13
88% formic	0	38	79	99	75	4
Bases						
30% NAOH	89	7	100	100	100	63
28% NH_4OH	85	0	100	100	100	99
n-butylamine	91	0	0	0	49	100
Aniline	85	0	0	0	96	200
Chlorinated Solvents						
Chloroform	57	0	0	0	87	100
Carbon tetrachloride	76	0	17	0	100	–
Chlorobenzene	73	0	0	0	100	100
Hydrocarbons						
Heptane	84	100	100	36	91	98
Xylene	91	0	0	0	100	100
Cyclohexane	90	75	99	0	100	–
Diesel fuel	87	100	100	36	100	–
Gasoline	80	99	100	0	100	–
Stoddard solvent	86	100	100	0	100	–
Dowtherm	89	0	0	0	100	–
Other Organic Solvents						
Methyl ethyl ketone	87	0	0	0	100	100
Cyclohexanol	84	74	95	27	100	96
Butyl alcohol	87	94	100	84	100	100
Phenol	0	0	0	0	100	100
Cellosolve	81	78	0	47	89	–
Dioxane	96	0	0	0	88	100
Nitrobenzene	100	0	0	0	100	100
Acetonitrile	93	25	0	69	96	100
N-N-dimethylformamide	95	0	0	–	100	94
Ethyl acetate	89	0	0	0	100	100
Aqueous Salt Solutions, and H_2O						
Water	66	100	100	100	100	–
Sodium chloride, 10%	94	100	100	100	100	–
Sodium hypochlorite, 10%	44	100	100	100	84	85
Aluminum sulfate (alum), 10%	33	100	100	100	100	–
Ammonium sulfate, 10%	62	100	100	100	100	–
Ammonium nitrate, 10%	47	100	100	100	100	–

Table 3.3-26. Neat resin chemical resistance (room temperature exposures) [69–71].

Chemical	Polyetherether ketone PEEK		Polyphenylsulfone Radel®		Polyphenylene Sulfide PPS (Ryton®)	
	Stress psi	Exposure* Time	Stress psi	Rupture Time	Stress psi	Exposure Time
Dichloromethane	1,000	1 week	500	3 min	1,000	1 week
			1,000	10 sec		
MEK	1,000	1 week	500	1 sec	1,000	1 week
T-5351 AL[1]	1,000	1 week	500	4 min	1,000	1 week
MIL-H-5606C[2]	1,000	1 week	500	No failure	1,000	1 week
Skydrol 500 B[3]	1,000	1 week	500	30 min	1,000	1 week
Acetone	1,000	1 week	500	20 min	1,000	1 week
Trichloroethylene	–	–	500	1 sec	–	–

*No failures.
[1]Air Force paint stripper.
[2]Air Force hydraulic fluid.
[3]Commercial aircraft hydraulic fluid.

In the case of polyetherether ketone (PEEK)—where rigid rings are connected by fairly chemically inert groups ($-O-$ and $-C-$) to give a highly crystalline,

$$\parallel$$
$$O$$

high melting point polymer—the chemical resistance is outstanding. According to ICI data, PEEK is soluble only in concentrated sulfuric acid. The performance of PEEK, PPS, and polyphenylsulfone in a number of solvents is compared in Table 3.3-26. After one-week immersions under stress, PEEK and PPS were unaffected by all the solvents, in contrast to polyphenylsulfone (Radel®), which failed in all but one environment. Similar comparisons had previously shown Radel® to have better chemical resistance than polysulfone (Udel®, P-1700) and polycarbonate [70]. Hartness [70] concluded from evaluations of several thermoplastic candidates that only semicrystalline polymers could meet the chemical resistance requirements for an advanced composite.

Moisture absorption for the high-performance thermoplastic matrix candidates is generally low compared to that of the MY-720-based structural epoxy incumbents. Typical moisture absorption for thermoplastics (Table 3.3-23) is less than 2% at equilibrium in water immersion or 100% RH air. The corresponding value for the neat epoxy is 5–5.5%. In both cases the equilibrium amount of moisture in the resin depends primarily on the relative humidity to which the sample is exposed. The effect of temperature is to influence the rate

at which equilibrium is reached. In Figure 3.3-25 the moisture absorption vs. square root of time curves for two thermoplastic resins are compared to that of a structural epoxy (tetrafunctional epoxy cured with diaminodiphenyl sulfone).

The effect of absorbed water on neat resin properties is much less, generally, for the thermoplastics than for the structural epoxies. This difference can be seen in Figure 3.3-26, where tensile stress-strain curves are plotted for polyphenylsulfone (Radel®) and 3501-6 epoxy under hot/wet conditions. The tensile strength of 3501-6 epoxy, measured at 150°C after conditioning at 100% RH, is only 6% of the value measured at 150°C on dry specimens; the corresponding value for polyphenylsulfone (at 177°C) is about 66%. This hot/wet performance advantage for thermoplastics in neat resin form also holds for the composite, as will be shown in a subsequent subsection.

In some applications the thermal stability of the resin under dry conditions may be of more interest than hot/wet behavior; in these cases the high-performance thermoplastics are generally excellent—several grades have been rated for continuous use in the 150–200°C range. By thermogravimetric analyses (TGA) the polysulfones, as a group, show little or no weight loss to 400°C in air or nitrogen; the maximum rate of degradation occurs in the 450–550°C range under both oxidative and anaerobic conditions [72]. On the other hand, amine-cured DGEBA epoxies degrade at significant rates in the 300–350°C range, even under anaerobic conditions, due to nonoxidative reactions of the amino-alcohol cure linkage, which is reported to be less stable than the substituted glycerol repeat unit. Reductions in cross-link density would be expected at temperatures lower than those required for significant weight loss; they have been reported to occur as low as 240°C [73]. These differences in thermal stability as measured by TGA are evidenced in longer exposures at lower temperatures; for example, Novak [74], in a comparison of two sulfone polymers and an epoxy, found 18% tensile strength retention for the epoxy vs. 60–100% for the two polysulfones after exposure at 177°C for 2,400 hours.

Continuous use temperature for reinforced PEEK is reported by the manufacturer to be 220–240°C, but no TGA data are provided [75]. Most of the other thermoplastic resins of interest (PPS, Torlon®, Ultem®) also have excellent thermal stability.

Smoke and Flammability Characteristics. The smoke characteristics of interior parts are of increasing concern to aircraft companies. Incumbent matrix resins are mostly 250°F-curing fire retardant epoxies, which generate more smoke than the unmodified resins. There is much current interest in phenolics for interior applications because these matrix resins offer both low flammability and low smoke but lead to more difficult processing. Most of the high-performance thermoplastics being considered for matrix resins have relatively good smoke characteristics when compared to other thermoplastics or to epoxy and polyester thermosets. The best of them (e.g., polyethersulfone) approach the phenolics and bismaleimides, which are the resins of choice when low smoke is the prime consideration. Since smoke generation is highly dependent on conditions of combustion, standard tests are necessary for meaningful comparisons; the most widely used method is the NBS smoke chamber test [76,77], which measures the density of smoke evolved from a sample under a set of standard conditions. A plot of smoke density, D_s, vs. time is shown in Figure 3.3-27 for several thermoplastic resins and thermoset laminates exposed in an NBS chamber. Both the total amount of smoke produced and the rate at which it reaches a maximum are important. Smoke densities at 1.5 and 4.0 minutes, for example, are called out in many aircraft specifications. The high-performance thermoplastic resins in Figure 3.3-27 develop a relatively small amount of smoke slowly, while ABS develops a large amount of smoke very quickly. The non-brominated epoxy lami-

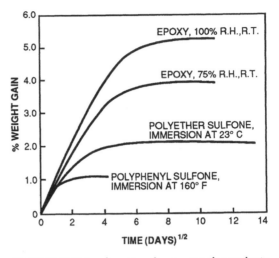

FIGURE 3.3-25. Water absorption of neat resins, thermoplastic vs. tetrafunctional epoxy.

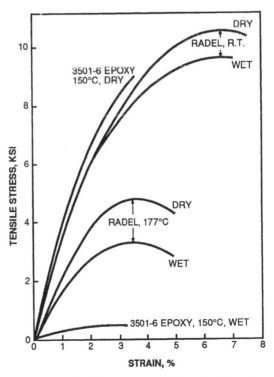

FIGURE 3.3-26. Tensile stress vs. strain for neat resins under hot/wet conditions.

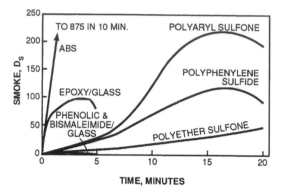

FIGURE 3.3-27. Smoke concentration histories for thermoplastics and thermosets in NBS chamber (neat resins except where noted otherwise; exposure: 2.5 watts/cm², flaming).

nate also generates its smoke very quickly and is clearly much worse than the phenolic and bismaleimide laminates. The resin remains the most important source of smoke when the reinforcement is graphite or Kevlar.

Oxygen index (O.I.) values for several thermoplastics are compared to those for neat epoxy resins and thermoset/glass laminates in Table 3.3-27. The O.I. values have been correlated with char yields [78–80]. In general, a high char yield correlates with a high O.I. value. All of the thermoplastic matrix resin candidates in the

Table 3.3-27. *Oxygen index for matrix resins and composites [78,79].*

Resin	O.I. at 23°C
Polyetherimide (Ultem®)	47
Polyphenylene sulfide (Ryton®)	44
Polyamide-imide (Torlon®)	43
Polyethersulfone (Victrex®)	38
Polyphenylsulfone (Radel® R-5000)	38
Polyetherether ketone (Victrex®)	35
Bismaleimide	35
Polysulfone (Udel® P-1700)	30
Phenolic	25
Epoxy	
DGEBA + MDA	27
+ TETA	23
+ NMA	22
+ HHPA	20
MY–720 + DDS	22
Bismaleimide (40 w/o)/S-181 glass fabric	60
Phenolic (40 w/o)/S-181 glass fabric	57
Epoxy (40 w/o)/S-181 glass fabric	27

table have higher O.I. values than any of the epoxy/hardener combinations, including the base resin for current state-of-the-art structural epoxies, MY-720. Vinyl polymers, as a group, give low char yields and low O.I. values [80].

Electrical Properties. Electrical properties of composites depend strongly on the matrix resin; for dielectric applications, such as radomes, the dielectric constant and dissipation factor (or loss tangent) (Volume 2) are important, as is their variation with temperature, frequency, and moisture content. Dielectric properties of several glass filled thermoplastics are shown in Table 3.3-28, measured at a millimeter radar frequency (35 GHz) under as-received and humidity-exposed conditions. The major effect of the humidity exposure is an increase in the dissipation factor.

The dielectric properties for the thermoplastics in Table 3.3-23 are generally equivalent to or better than those of DGEBA epoxies at the same frequencies. At 10^6 Hz, the epoxies have dielectric constants from 2.4 for the best anhydride-cured system to about 3.9 for the same resin cured with an aromatic amine; all of the thermoplastics in Table 3.3-23 fall in this range. The corresponding dissipation factors are 0.008 and 0.038 for anhydride- and amine-cured epoxies; several of the thermoplastics have better (lower) dissipation factors.

Polysulfone reinforced with Kevlar 49 or glass fabrics showed dielectric constant performance at radar frequency (X-band) equivalent to that of epoxies at both room temperature and 250°F; the loss tangent performance was superior [82].

Processing

The use of thermoplastics as matrix resins replacing epoxies requires a fundamental change in processing, since the emphasis shifts from cure chemistry to manipulation of a high molecular weight polymer to form a well-consolidated shape. The use of a preformed polymer vs. a mixture of monomeric or oligomeric reactants has already been mentioned as an advantage for thermoplastics; however, the advantage is accompanied by some potential disadvantages.

Much higher processing temperatures are needed to produce a part with a given use temperature. This requirement can be seen in Table 3.3-29, where the melt processing temperatures for a number of thermoplastics are compared to the heat deflection temperature (HDT) of the unreinforced polymer. Note that in most cases

Table 3.3-28. Electrical properties of some thermoplastic composites before and after humidity exposure [81].

Resin	Reinforcement	Dielectric Constant		Loss Tangent	
		As-received	Exposed	As-received	Exposed
Polyethersulfone	40% glass	3.8	3.9	0.012	0.019
Polysulfone	40% glass	3.8	3.9	0.012	0.019
Polyphenylene sulfide	40% glass	3.8	3.9	0.002	0.010
Polyamide-imide	30% glass	4.0	4.2	0.023	0.036
Nylon	40% glass	3.8	4.0	0.012	0.035
Polyester	– glass	3.4	3.6	0.007	0.014
Polycarbonate	40% glass	3.6	3.7	0.011	0.014
Polybutadiene	– glass	4.2	4.6	0.010	0.043
Fluorocarbon	20% glass	2.2	2.6	0.003	0.020

Note: "Exposed" = Humidity exposure according to MIL-STD-810, Procedure I cycled between RT & 150°F (65°C) at 85-95% RH for 240 hours.
Frequency: 35 GHz.

processing temperatures start well above 300°C and in some cases extend to 400°C or above; the resulting heat deflection temperatures are generally no higher than those of the multifunctional epoxies cured (and/or postcured) at 175–225°C. The higher temperature requirements for thermoplastics do not necessarily mean higher overall energy requirements, since cycle times can be much shorter (one second to two hours). Higher processing temperatures also lead to higher tooling costs and, therefore, to more capital investment.

In addition to these economic considerations, there are several technical concerns in processing thermoplastics. A major consideration is achieving good wet-out of the individual filaments in a fiber bundle instead of merely encapsulating the bundle, which can easily occur in melt impregnation. The melt impregnation process must also be carried out under time/temperature and environmental conditions which prevent degradation of the resin or fiber. In solvent impregnation, a major concern is removal of the solvent prior to final consolidation of the part. Incomplete removal leads to low mechanical properties and is most often a problem with highly polar, high-boiling solvents necessary for some resins. May and Goad [83], for example, reported a 1.8-fold increase in 0° compression strength of AS/polysulfone by changing the impregnating solvent from NMP (N-methyl pyrollidone) to methylene chloride. Unfortunately, resins soluble in easy-to-remove solvents have (not too surprisingly) shown poor solvent resistance, so the trend is to use non-solvent impregnation methods (e.g., film and fabric or melt impregnation).

Table 3.3-29. Processing temperatures for thermoplastic resins (unreinforced).

Polymer		HDT, °C	Melt Temperature, °C (for injection molding)
Polybutylene terephthalate		56	238–250
Polyphenylene sulfide	–Ryton®	137	315–340
Polyetherether ketone	–Victrex® PEEK	140	360–400
Polycarbonate	–Lexan® –Merlon®	145	280–340
Polyarylate	–Ardel®	174	340–390
Polysulfone	–Udel®	174	340–390
Polyetherimide	–Ultem®	200	340–425
Polyethersulfone	–Victrex® PES	203	350–400
Polyphenylsulfone	–Radel®	204	370–425
Polyamide-imide	–Torlon®	274	330–360

Table 3.3-30. Composite fabrication conditions (graphite reinforcement).

Resin	Impregnation Method	Consolidation Conditions			Reference
		Temperature °C	Pressure psi	Time Minutes	
Polysulfone	Tape, Prepreg from MC	315	500	5	[90]
Polyethersulfone	Tape, Prepreg from DMF	315	2,000	60	[90]
Astrel® 360	Tape, Prepreg from DMF	370	2,000	60	[90]
Radel® 5000	Film + Fabric	370	2,000	–	[69]
PPS	Film + Fabric	315	100	30	[74]
PET, PBT	Film + Fabric	275	100	30	[91]
PEEK	Film + Fabric	400	200	120	[70]
PEEK	ICI Tape (melt)	380	150	2 seconds	[92]

MC = Methylene Chloride.
DMF = N,N-Dimethylformamide.

The three major approaches which have been used to combine resin and fiber for continuous fiber reinforced thermoplastics are as follows:

• solvent impregnation of tape or fabric
• interleaving of resin film and fabric (film stacking)
• melt impregnation, usually of tape

The solvent method used commercially to prepare polysulfone prepregs has been widely used in small-scale evaluation work. The melt method is apparently being used commercially by ICI to prepare PEEK/graphite tape. Interleaving of film and fabric appears to be most useful for small-scale evaluations, especially those in which the matrix resin is insoluble or soluble only in unfriendly solvents. It has been successfully used with polyphenylene sulfide, polyphenylsulfone, polyethersulfone, and PEEK [70,84].

If thermoplastics become widely accepted as matrix materials, it seems likely that, in addition to fabric or tape "prepregs" prepared from solvent or melt, some form of pre-consolidated blank (i.e., sheet stock) will also be supplied by the vendor, which, in the case of PEEK/graphite, is already being done. For such stock to be useful, it must be reformed into a useful shape without loss of the properties present after the initial consolidation.

In an early processing study, Hoggatt [85] demonstrated that polysulfone (P-1700) with either glass or graphite reinforcement could be post-formed from standard sheet stock into structural shapes (e.g., corrugated panels) without degrading the quality of the laminate; the resulting part quality was equivalent to

that of an epoxy control. Subsequent studies have confirmed and extended this work to other systems [86–89].

Optimum processing methods have not been established in most evaluations because of time constraints and the complexity of the problem; a typical evaluation involves preparation of laminates by only one method. With a given method, optimization studies have sometimes been carried out; Novak, for example, considered both impregnation and consolidation variables in a study [90] of polysulfone, polyethersulfone, and poly-arylsulfone as matrix resins for graphite. The consolidation conditions selected from this study are shown in Table 3.3-30 along with conditions for other resins from several additional studies.

An indication of the importance of fabrication conditions on shear strength of the composite can be seen in Table 3.3-31, where the short beam shear (SBS) strength for several fabrication routes is shown for a polysulfone/graphite tape [90]. The data illustrate the conditions which influence the interlaminar shear strength and also show the difficulty of pinpointing the critical fabrication variables which lead to good mechanical properties.

The 0° flexural strength of a graphite/polysulfone laminate was studied to establish an optimum "properties-pressure-temperature" zone for a cruise missile prototype composite wing in which polysulfone was used with Kevlar and S-glass reinforcement for the skins and with graphite (short-fiber, injection molded) for the frame [93]. For a given lay-up and dwell time at 600°F, a minimum pressure of 150 psi was required to obtain optimum flex strength. At a pressure of 100 psi, increasing consolidation temperature to 650°F was needed in order to be in the zone of optimum flex strength.

Thermoplastics are also suitable for use in other processes usually reserved for thermosets. Pultrusion of graphite/polysulfone and graphite/PEEK in tape and fabric form has been used to prepare flat panels, angle sections, and sandwich panels [87,94]. Product literature from ICI indicates that PEEK/graphite tapes can be used for filament winding or tape laying.

Thermoplastic fibers have been blended, "comingled" with graphite fibers, and then fused in a melt process. Preliminary results with a thermoplastic polyester (polybutylene terephthalate) indicate that good wetting is possible [95]. A melt impregnation process for producing thermoplastic composite rods reinforced with glass or Kevlar has also been used commercially. Matrix resins include PPS, PBT, 6 and 66 nylon, ABS, and polypropylene [95].

Performance of Thermoplastics vs. Epoxies in Composites

Most of the evaluation work on thermoplastics as matrix resins for continuous fiber composites has been in the context of aerospace applications, and the approach has been to compare them with incumbent epoxies for these uses. In addition to the processing studies already discussed, the effort has centered on establishing mechanical property equivalency at ambient and elevated temperatures, on defining hot/wet performance, and on determining chemical resistance of the thermoplastic matrix candidates to the specific solvents commonly found in aircraft applications. The resins which have received the most attention are PEEK, polysulfone, and other sulfone polymers (polyphenylsulfone, polyarylsulfone, polyethersulfone). Various other resins, including modified polysulfones and polyimides, are also being evaluated. A look at the HDT column in Table 3.3-23 will show why these resins were chosen. Polysulfone was the first commercially available thermoplastic resin which had the balance of mechanical properties and high T_g needed to compete with incumbent epoxies. Subsequent commercial introduction of PEEK and other polymers has raised the T_g and improved the solvent resistance vs. polysulfone. In this discussion, we will examine some of these resins, concentrating on those matrix-dominated properties which separate one matrix resin from another.

Polysulfone in Composites. The performance of polysulfone P-1700 (Udel®) has been compared to epoxies and to other thermoplastics, and it has been used in fabrication of aircraft parts for flight service evaluation [96]. The consensus from a number of evaluations [74,82,90] is that polysulfone is useful only below 300°F; at 350°F it has negligible load-carrying capability [74]. Polysulfone is therefore not a candidate for replacement of current structural epoxies based on MY-720. In addition to this high temperature limitation, polysulfone's other major drawback is its poor resistance to various aircraft fluids when compared to epoxies or even to most of the other thermoplastic candidates. Except for these two limitations, polysulfone composites with graphite, aramid, and glass fibers have generally performed well.

Graphite/polysulfone properties are summarized in Table 3.3-32 for S-181 fabric reinforcement: at 300°F the matrix-dominated property retentions are 54–69% (compression, flexural, shear strengths), whereas lon-

Table 3.3-31. Polysulfone (P-1700)/graphite prepreg tape processing [90].

Sample No.	Relative SBS	Prepreg				Consolidation		
		% Solids	Solvent	Drying		Temp. °C	Time Minutes	Pressure psi
1	100	10	MC	150°C, Air		315	60	500
2	98	10	MC	150°C, Vac.		315	5	500
3	91	10	MC	150°C, Air		315	30	500
4	83	8.7	MC + TCE	80°C, Vac.		270	5	5,000
5	59	8.7	MC + TCE	80°C, Vac.		270	5	1,000
6	59	10	MC	150°C, Air		315	5	500
7	43	10	MC	Air		270	5	1,000

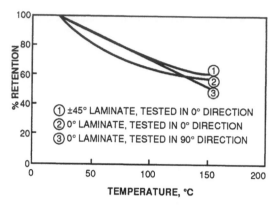

FIGURE 3.3-28. Percent retention of compressive strength vs. temperature, polysulfone/A-S graphite tape laminates (data from reference [86]).

gitudinal tensile strength and all moduli are 80–100% of room temperature values. Unidirectional and ±45° tape laminate data on AS/P-1700, plotted in Figure 3.3-28, show compression strength retentions at 300°F of 57% and 51% for longitudinal and transverse properties of unidirectional laminates, respectively—and 60% for ±45° specimens. Interlaminar shear strength results are shown in Figure 3.3-29 for graphite/P-1700 [86,90]. Flexural strength retentions for graphite/polysulfone [86,90] are shown in Figure 3.3-30, together with data for graphite/PEEK [97].

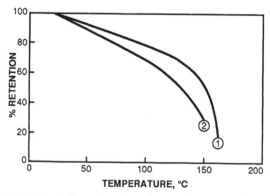

FIGURE 3.3-29. Retention of interlaminar shear strength vs. temperature for graphite/polysulfone (P-1700 polysulfone, unidirectional): (1) data from reference [90]; (2) data from reference [86].

Table 3.3-32. Graphite fabric/polysulfone P-1700 properties [86].

Tensile	
Strength, 10^3 psi	
R.T.	77.1
300°F	61.4
% Retention	80
Modulus, 10^6 psi	
R.T.	10.0
300°F	10.4
% Retention	>100
Compressive	
Strength, 10^3 psi	
R.T.	56.0
300°F	36.2
% Retention	65
Modulus, 10^6 psi	
R.T.	9.1
300°F	10.6
% Retention	>100
Flexural	
Strength, 10^3 psi	
R.T.	107
300°F	68.6
% Retention	64
Modulus, 10^6 psi	
R.T.	7.7
300°F	7.2
% Retention	94
Interlaminar Shear Strength, 10^3 psi	
R.T.	6.5
200°F	3.5
% Retention	54

Polysulfone and phenoxy glass fabric laminates were compared to an epoxy prepreg system (Narmco 551-181, BMS 8-79) by Hoggatt [85]. The study also includes other mechanical property data, including impact, creep, fatigue, thermal stability, flammability, environmental, and electrical property comparisons. Impact strength improvements of 30–40% over that of epoxy were found for polysulfone and phenoxy matrices. Other properties were also equivalent or better for polysulfone, and the overall conclusion was that none of the tests precluded its use in aircraft structures.

Kevlar 49/polysulfone laminates were evaluated by Hoggatt and Von Volkli [82], who found the reinforcement to be compatible with impregnation process, laminating, and postforming conditions needed for polysulfone—the major problem was low fiber-resin adhesion. Polysulfone was also evaluated as a matrix for hybrids of graphite/Kevlar, graphite/glass, and Kevlar/glass with good results [98].

Creep results for graphite/polysulfone ±45° laminates are shown in Figure 3.3-31 [74]. Also shown are results for epoxy matrix composites. Steady state creep rate is much lower for the epoxy than for the two thermoplastics, which is in agreement with a similar test on the neat resins.

Capped Polysulfone. Although several studies have rated polysulfone acceptable in terms of solvent resistance (when tested in certain solvents, unstressed, and at room temperature), its resistance to chlorinated solvents is poor (it is soluble in methylene chloride, for example). Consequently, a considerable amount of work has gone into modifying the polymer to reduce its solvent sensitivity while retaining property and processing advantages. Usually these efforts have involved "capping" the polymer with reactive end groups which serve as cross-linking sites during postcure. The objective is to provide sufficient cross-linking to improve solvent resistance while retaining thermoforming capability. The use of a norborene end cap on a polysulfone oligomer of appropriate molecular weight has been reported [88] to give a matrix whose graphite composites remained intact (with only slight swelling) after immersion in methylene chloride for two months at room temperature—a control graphite/P-1700 composite had ply separation in three hours. In addition, shear strength for the modified system (NTS 20-1) was equivalent to the control, and reformability (thermoforming) was judged equivalent to the P-1700 control. The same approach to improving chemical resistance has been used with other end groups [99].

Other Polysulfones. In addition to polysulfone (Udel® P-1700) itself, three other polysulfones have been evaluated in composites: polyarylsulfone (Astrel® 360), polyphenylsulfone (Radel® 5000, Union Carbide), and polyethersulfone (Victrex® PES, ICI). Astrel 360, originally introduced by 3M Co., is no longer commercially available.

Radel polyphenylsulfone and PES polyethersulfone were evaluated by Hill et al. [100] as matrix resin candidates to provide improved environmental resistance over polysulfone (Udel® P-1700) in graphite fabric laminates for aircraft use. Baseline mechanical properties from this study are summarized in Table 3.3-33; the fabric style and the prepregging solvent were identical, and the consolidations were similar for this comparison. Compressive strength retention for Radel at 350–

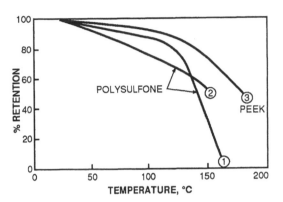

FIGURE 3.3-30. Percent retention of flexural strength vs. temperature for graphite/polysulfone and graphite/PEEK (unidirectional laminates): (1), (2) and (3) denote data from references [90], [86], and [97], respectively.

400°F is somewhat higher than for PES, and both show better property retentions in this temperature range than polysulfone because of their 25–30°C higher HDT. Better solvent resistance was found for Radel vs. PES in a series of aircraft liquids; in exposures under stress (60% of ult.), PES specimens (flex bars) delaminated in all environments, whereas Radel delaminated in only one hydraulic fluid and retained over 85% of its initial strength in the others. In another study, Husman and Hartness [69] evaluated graphite/polyphenylsulfone fabric laminates made by film stacking under several en-

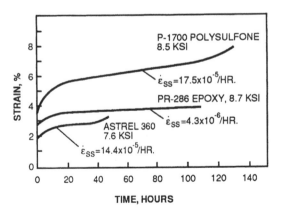

FIGURE 3.3-31. Creep of ±45° T-300 laminates with epoxy and thermoplastic matrices at 121°C [74].

Table 3.3-33. Mechanical properties of graphite fabric laminates in polyethersulfone and polyphenylsulfone matrices (8-H satin fabric; solvent prepreg from DMF) [100].

	Temp. °F	Polyethersulfone		Polyphenylsulfone	
		PES 300 P		Radel® 5010	
			% Retention		% Retention
Tensile	75	70.7	–	63.6	–
Strength,	250	57.1	81	59.3	93
10³ psi	350	49.8	70	51.4	81
	400	47.5	67	35.2	55
Tensile	75	8.2	–	8.3	–
Modulus,	250	7.1	87	7.6	92
10⁶ psi	350	7.1	87	6.3	76
	400	5.7	70	4.9	59
Compression	75	70.8	–	60.5	–
Strength,	250	53.7	76	53.9	89
10³ psi	350	47.5	67	47.5	79
	400	28.2	40	34.8	58
Compression	75	9.3	–	8.1	–
Modulus,	250	8.7	94	7.3	90
10⁶ psi	350	8.7	94	5.9	73
	400	8.8	95	4.1	51
Flexural	75	89.0	–	102.3	–
Strength,	250	56.3	63	84.9	83
10³ psi	350	53.9	61	66.7	65
	400	34.9	39	39.9	39
Flexural	75	8.6	–	8.1	–
Modulus,	250	5.9	69	7.7	95
10⁶ psi	350	5.4	63	5.1	63
	400	4.7	55	4.1	51
Short Beam	75	6.8	–	7.7	–
Shear Str.,	250	4.4	65	5.4	70
10³ psi	350	–	–	–	–
	400	2.5	37	2.4	31
Consolidation		640°F/2 hr. +640°F/0.75 hr./200 psi +		745°F/0.5 hr. 745°F/0.5 hr./200 psi	

vironmental conditions and concluded that the resin offered excellent moisture resistance, thermal stability, and mechanical properties for use to 350°F; chemical resistance to solvents was the major concern. A summary of the hot/wet properties is presented in Table 3.3-34. This study also demonstrated thermoforming of graphite/polyphenylsulfone to be relatively easy at 450°F.

Polyetherether Ketone (PEEK). Since its introduction in 1981, PEEK has generated considerable interest as a high-performance matrix resin candidate. PEEK/graphite composites prepared by the film and fabric

technique were evaluated by Hartness [70], who concluded that PEEK has excellent potential as a matrix resin based on its good hot/wet mechanical properties, room temperature creep, and toughness; neat resin solvent resistance was also very good. The hot/wet data from this study are summarized in Table 3.3-35, and the flex strength data are plotted in Figure 3.3-32. Under dry conditions the epoxy clearly maintains stiffness and strength better at high temperatures, but the situation is reversed under wet conditions because the stiffness and strength vs. temperature curves for PEEK are the same wet or dry, whereas the epoxy curves show large decreases for moisture conditioned samples. At 177°C the flex strength of the PEEK/graphite laminate is about 1.8 times that of the epoxy/graphite after moisture conditioning.

The toughness comparisons in the Hartness study (Table 3.3-36) show the critical delamination strain energy release rate (G_{IC}) values for PEEK/graphite to be 8.5 to 10 times those of the epoxy control and about twice those of a polysulfone control. Higher values in this test indicate more resistance to delamination. Creep resistance of graphite/PEEK, measured on ±45° laminates, was superior to that of graphite/epoxy in room temperature tests; the creep curves are shown in Figure 3.3-33.

Properties of PEEK reinforced with unidirectional graphite fiber reported by Cooke [97] are tabulated in Table 3.3-37. The high retention of shear strength at 212°F after 10 days in boiling water and the exceptionally good transverse tensile strength (14.5 ksi) are noteworthy results attributable primarily to the matrix and the good adhesion to the fibers. The transverse tensile strength value is 1.6 to 2.3 times that reported for structural epoxies from three of the major suppliers [101].

PEEK is also available from the manufacturer, ICI, as the matrix resin for a graphite prepreg which is designated APC-2 (aromatic polymer composite). The impregnation method is proprietary; however, it is thought

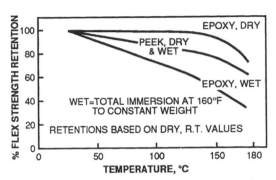

FIGURE 3.3-32. Flexural strength vs. temperature for graphite fabric reinforced PEEK vs. epoxy (data from Hartness [70].

to be a melt process. The product forms available are tape or sheet stock. Unidirectional tape is available in widths up to about 6 inches, the thickness is about 0.005 inch, and the fiber volume is 60%. The prepreg tape is said to be fully wet-out and to contain less than 0.5% voids.

APC-2 prepreg tape can be plied as desired and consolidated at 380°C at a pressure of 150 psi and times as short as 1–2 seconds. No chemical reactions occur during the consolidation. Sheet stock can be formed in the 360–400°C range (PEEK melts at 340°C) by heating to temperature and then applying pressure rapidly in a

Table 3.3-34. Properties of Radel® graphite/polyphenylsulfone laminates under hot/wet conditions (T-300, 8-HS fabric) [69].

Sample Condition	Flexural Strength 10³ psi	Flexural Modulus 10⁶ psi	SBS Strength 10³ psi	Tensile Strength 10³ psi Direction Warp	±45°	In-Plane Shear Strength 10³ psi	Modulus 10³ psi
Dry							
RT	79.3	4.16	9.03	36.4	18.1	9.07	479
350°F	46.5	3.10	6.66	39.1	11.4	5.71	507
% Retention	59	75	74	>100	63	63	>100
Wet							
RT	61.3	3.53	8.54				
% of Dry Value	77	85	95				
350°F	45.5	3.65	5.43				
% of Dry RT Value	57	88	60				
% of Wet RT Value	74	>100	64				
% of Dry 350° Value	98	>100	82				

Notes
Fiber Volume: 40%
"Wet" = conditioned by immersion in 160°F H₂O to 0.46–0.58% weight gain.
In-plane shear results from ±45° tensile tests.

Table 3.3-35. Comparison of hot/wet performance of PEEK and 350°F epoxy fabric laminates [70].

	Flexural Strength, 10^3 psi		Flexural Modulus, 10^6 psi	
	PEEK	**Epoxy**	**PEEK**	**Epoxy**
Dry				
RT	102	113	7.43	7.85
121°C (250°F)	85	108	6.87	7.64
% of RT Value	83	96	92	97
177°C (350°F)	59	80	5.99	7.88
% of RT Value	58	71	81	100
Wet				
RT	102	120	7.43	8.51
% of Dry RT Value	100	106	100	108
121°C (250°F)	83.4	71.4	6.87	7.99
% of Dry RT Value	82	63	92	102
% of Wet RT Value	82	60	92	94
% of Dry 121°C Value	98	66	100	100
177°C (350°F)	60.0	34.0	5.99	5.78
% of Dry RT Value	59	30	81	74
% of Wet RT Value	59	28	81	68
% of Dry 177°C Value	102	43	100	73
% H_2O (Equil.)	0.42	1.34	0.42	1.34
Fiber Volume, %	57	62	57	62

Moisture Conditioning: Total immersion at 160°F to constant weight.
Fiber: T-300, 8-hardness satin weave; used in both PEEK and epoxy laminates.
Epoxy Prepreg: Fiberite HMF-133-76, 350°F cure.

Table 3.3-36. Comparison of graphite fiber PEEK and epoxy composites [70].

±45° **Properties** (**In-Plane Shear**)	**T-300/Epoxy 8-HS**	**T-300/PEEK 8-HS**
Tensile Strength, ksi	26.3	26.5
Tensile Modulus, 10^6 psi	2.24	3.41
Shear Strength, ksi	13.1	13.2
Shear Strain, ult., %	3.27	7.38
Poisson's Ratio	0.773	0.606

Toughness	**Critical Strain Energy Release Rate (G_{1c}) in-lb/in²**	
Fabric Reinforcement	G_{1c}	**Relative**
Epoxy	1.34	1
PEEK	11.4	8.5
Unidirectional		
Epoxy	0.8	1
PEEK	8.0	10
Polysulfone	3.7	4.6

Note: Epoxy is 350°F state-of-the-art commercial prepreg designated HMF-133-76 (Fiberite).

Table 3.3-37. Properties of Grafil XA-S graphite fiber/PEEK composites [97].

Tensile	
0° Strength, 10^3 psi	280–290
0° Modulus, 10^6 psi	18.5
Flexural	
0° Strength	
RT, 10^3 psi	240–290
250°F	230
% Retention	79–96
350°F, 10^3 psi	130
% Retention	45–54
0° Modulus	
RT, 10^3 psi	17.5–19.6
250°F	19.0
% Retention	97–100
350°F	16.8
% Retention	86–96
90° Strength, 10^3 psi	14.5
90° Modulus, 10^6 psi	1.5
Interlaminar Shear Strength (S/D = 5/1)	
RT, 10^3 psi	15.4
212°F, 10^3 psi	14.5
% Retention	94
212°F, After 10-Day Water Boil	13.8
% of RT Value	90
% of 212°F Value	95

cold tool. ICI recommends pressures from 80–1,000 psi depending on the forming process:

- hydraulic diaphragm – 1,000 psi
- pneumatic diaphragm – 80 psi
- matched-die molding – 150 psi

A typical overall cycle (heating, forming, cooling) for these methods is about two minutes.

One of the often cited advantages of thermoplastics as matrix resins, remolding of scrap, has been documented by ICI in a comparison of a standard graphite/PEEK injection molding compound with a blend of reclaimed (scrap) graphite/PEEK with additional PEEK resin. Results, shown in Table 3.3-38, indicate that both stiffness and strength are better for the reclaimed material than for the standard molding compound.

Some typical property data (from ICI) are summarized in Table 3.3-39. Properties are, in general, impressive: the shear strength value (15 ksi) supports the claim of low void content (<0.5%), and the 0° modulus value indicates good translation of fiber properties to the composite. Stiffness retention as a function of temperature is shown in Figure 3.3-34: the flexural modulus retention is about 88% at 177°C (350°F) and 50% at 290°C.

Impact properties of graphite/PEEK are reported good at room temperature; at −50°C impact strength is 90% of the room temperature value. Graphite/PEEK, impacted in a falling weight impact test and then tested in compression, retained about half its initial strain to failure (1.1%) after absorbing 2,000 in-lb per inch of thickness. By contrast, a state-of-the-art graphite/epoxy loses about half its compressive strain-to-failure after absorbing about 20% as much impact energy [102].

Other Resins. Polyphenylene sulfide (Ryton®, PPS) was examined as a matrix resin with graphite by Hartness [71], who concluded it had excellent potential for use to at least 300°F. Chemical resistance, as expected from neat resin studies, was excellent even in methylene chloride or methyl ethyl ketone. After moisture conditioning to equilibrium at 160°F (0.6% weight gain), the flex strength retentions at 250°F and 350°F were 46% and 33% of the dry, room temperature values, which suggests that an upper use temperature of 300°F may be optimistic. Other results support such a conclusion: Sheppard and House [103] reported flex strength retention of 40% and compressive strength retention of 35% at 300°F for dry specimens. The latter value is consistent with the compressive yield strength retention of 30% reported for a crystalline, glass filled, injection molded material at 300°F [104].

Both polybutylene terephthalate (PBT) and polyethylene terephthalate (PET) have been evaluated as matrix resins for graphite fiber composites (see Table 3.3-40). Sheppard and House [103] obtained poor elevated temperature properties (50% reduction in flex strength at

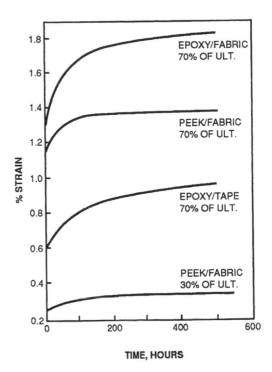

FIGURE 3.3-33. Creep of graphite/PEEK and graphite/epoxy laminates at room temperature, T-300 fabric (8-HS) and tape, ±45°.

Table 3.3-38. Comparison of graphite/PEEK standard molding compound with APC-1/PEEK blend.

	Wt % Fiber	Flexural Strength 10³ psi	Flexural Modulus 10⁶ psi
Standard Unfilled PEEK	0	23	0.60
Standard Graphite/PEEK Molding Compound	30	50	3.0
APC-1/PEEK (reclaimed APC-1)	25	53	3.2
APC-1/PEEK (reclaimed APC-1)	50	54	5.9

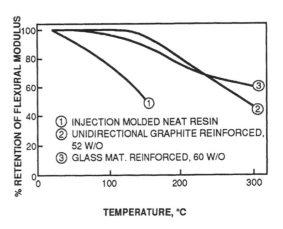

FIGURE 3.3-34. Percent retention of flexural modulus vs. temperature for PEEK resin and laminates.

Table 3.3-39. Properties of unidirectional graphite/
PEEK (all data from ICI).

Fiber Volume, %	52
Density g/cm³	1.60
Flexural	
0° Modulus, 10⁶ psi	17.4
0° Strength, 10³ psi	242
0° Strain at Failure, %	1.4
90° Modulus, 10⁶ psi	1.5
90° Strength, 10³ psi	16
90° Strain at Failure, %	1.4 (yielding)
0°/±45°/90° Modulus, 10⁶ psi	6.8
0°/±45°/90° Strength, 10³ psi	102
0°/±45°/90°Strain at Failure, %	1.4
Interlaminar Shear Strength, 10³ psi (ASTM D-2344, S/D = 5/1)	15
Shear Modulus, 10³ psi	600
Coefficient Linear Thermal Expansion (23–140°C)	
0° Direction, PPM/°C	2.5
90° Direction, PPM/°C	31
Water Absorption, equil. at 95°C, %	0.25

200°F) on graphite/PBT unidirectional laminates and omitted PBT from further study. Both PET and PBT films were used with graphite fabric to prepare laminates in another study [91]. Solvent resistance was excellent under stress in seven environments, including methylene chloride and acetone. Although no hot/wet tests were made, both resins were thought to be susceptible to degradation by moisture under hot conditions based on the data of Knight [72], which predicted, for PBT, 50% loss in tensile strength after 3–4 years at 50°C/100% RH and loss in toughness and impact much earlier than the tensile half-life time. The flexural properties of dry PET and PBT/graphite laminates are plotted vs. temperature in Figure 3.3-35. Flexural strength retention is similar for both (38–42% at 350°F).

Poly(amide-imide) (Torlon®) was selected for evaluation by Hoggatt and Von Volkli [82] as one of a number of commercially available polymers that has potential for higher use temperature than polysulfone (300°F), is processable, and is amenable to prepreg manufacture. It was subsequently eliminated from the study because of less favorable performance vs. other candidates. In a later study, graphite/Torlon laminates exhibited excellent high temperature property retention (91–100% retention of flex strength and modulus at 300°F), but further work was postponed because of the difficulty of forming the laminates into usable shapes [105].

Several other resins have been considered as matrix candidates but have been rejected at an early stage, usually because of processing difficulties or solvent resistance. Ardel (aromatic polyester), for example, was eliminated [100] because of its solubility in methylene chloride. A thermoplastic polyimide (XU-218 from Ciba-Geigy) was eliminated from a five-resin study [88] after processing attempts led to low mechanical properties and chemical resistance tests indicated poor resistance to several solvents. A new thermoplastic polyetherimide (Ultem® from G.E.) has excellent properties and a high HDT but is amorphous and known to be susceptible to solvent stress crazing in several solvents [106]. No data are available on its performance as a matrix for continuous fiber reinforcement. Various other thermoplastic polyimides have been synthesized and characterized [107,108] on a lab scale, but they are not widely available.

Adhesive Bonding of Thermoplastic Composites

Since many of the thermoplastic matrix resins are also good hot-melt adhesives, they are useful in various bonding operations. Graphite/polysulfone can be bonded to itself with a polysulfone film to give tensile failure in the composites at 175 ksi and no bond failure; when the specimen was changed to induce failure in the

Table 3.3-40. Physical properties of graphite tape/
thermoplastic polyester composites, Celion® 6000
graphite/polybutylene terephthalate (PBT).

	Boeing Data $V_f = 60$	Celanese Data	
		$V_f = 50$	$V_f = 62$
Tensile strength, 10³ psi	—	160–200	190–240
Tensile modulus, 10⁶ psi	—	15.0–16.5	18.5–20.5
Flexural strength, 10³ psi			
RT	173	220–230*	270–285*
180°F	90.1	—	—
200°F	83.9	—	—
Flexural modulus, 10⁶ psi			
RT	13.9	14.5–16.5*	18.0–20.2*
180°F	12.7	—	—
200°F	12.7	—	—
Short beam shear str., 10³ psi			
RT	11.1	—	—
180°F	6.55	—	—
200°F	6.20	—	—

*4-pt. bending.

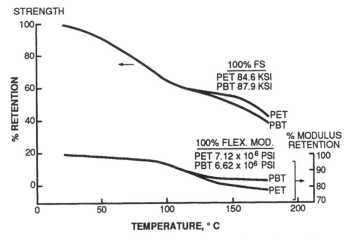

FIGURE 3.3-35. Flexural properties vs. temperature for thermoplastic polyesters with graphite reinforcement (T-300 fiber, 8-HS fabric; 55 v/o for PET, 60 v/o for PBT). Data from reference [91].

joint, lap shear strength was over 8,000 psi. The bond in this test was a simple one-inch lap joint formed at 450°/5 min/50 psi. Bonding of graphite/polysulfone to stainless steel with both epoxy and polysulfone adhesives gave equivalent bond strengths both as-fabricated and after 1,000 hours heat aging at 250°F [82]. PEEK is also a good hot-melt adhesive [109].

3.3.5 Polyimides

Introduction

Polyimides are high-performance polymers noted primarily for their high temperature capabilities, good mechanical properties, good chemical resistance, and low smoke characteristics. The first commercially available polyimides, introduced in 1962 by the DuPont Company, were thermoplastics with exceptionally good high temperature properties but of limited value as matrix resins because of their intractability. Subsequent work has led to the introduction of polyimides which maintain most of their desirable high temperature properties and offer much improved processability [110]. The major route to improving processability has been to convert the polyimides to thermosetting polymers by incorporating into the backbone some non-polyimide

functionality which can be used as a site for an addition-type cross-linking reaction as a final step in composite fabrication. A second route to improving processability has been to change the backbone structure to make the fully imidized thermoplastic polyimide more tractable in order to allow melt processing. The performance of these various polyimide types will be discussed in this subsection.

Although polyimides offer desirable characteristics for many applications, they remain in the category of low volume, high cost, specialty resins when compared to the major matrix resins (polyesters, epoxies) used in composites. Although polyimide sales figures are not reported separately by the SPI, other estimates [111] place total sales of polyimides for reinforced plastics in the 1–3 million pound/year range (vs. 10^9 pound/year for polyester and epoxies).

Condensation Polyimides

Polyimides are condensation products of diamines with aromatic tetracarboxylic acids, typically used as the dianhydride or the diester. Diamines may be aliphatic or aromatic [112], but most commercial polyimides are based on aromatic diamines to achieve maximum thermal stability. The preparation of polyimides, shown in Figure 3.3-36, proceeds in two stages, initially

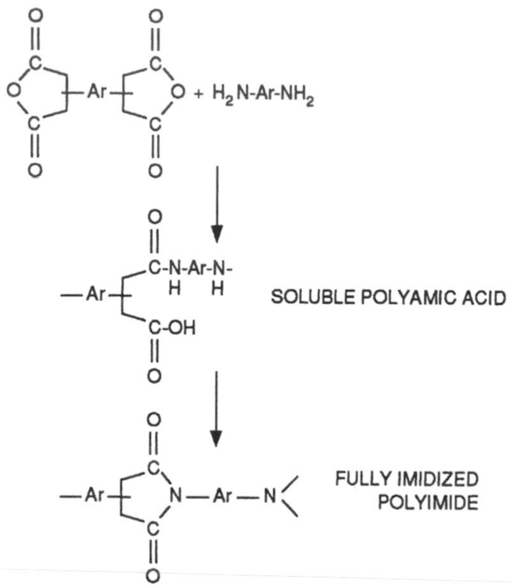

FIGURE 3.3-36. Condensation polyimide chemistry.

forming a soluble polyamic acid by reaction of diamine and dianhydride; this step, usually carried out at relatively low temperature (<100°C), determines molecular weight of the polymer. The second step, cyclization of the polyamic acid to form the polyimide ring, requires higher temperatures and liberates water as a by-product of the ring closure. Since complete imidization must be achieved to obtain optimum properties, water is evolved even in the last stages of composite formation and is one cause of porosity in the finished parts when the matrix resin is a condensation polyimide.

The most important ingredients used in condensation polyimides are shown in Figure 3.3-37. Pyromellitic dianhydride and benzophenone tetracarboxylic dianhydride are two dianhydrides of commercial importance which are typically used in combination with any of several aromatic diamines, some of which are shown in Figure 3.3-37. Commercial examples of polyimides from these ingredients include Skybond resins (Monsanto), BTDA/DPE and Kapton film (DuPont) from PMDA/DPE, and Pyralin coatings (DuPont). The Skybond and Pyralin systems are available only as "precursor" solutions of the intermediate polyamic acids (Figure 3.3-36) in a solvent which is usually polar, high boiling, and thus difficult to remove during laminate

preparation. The precursor solutions are typically used to form a prepreg fabric or tape, analogous to epoxy prepregs but much more difficult to convert to a void-free laminate because of the solvent and the by-products from the condensation cure.

Addition Types

Since the processability difficulties with condensation polyimides are due in large part to the need for removal of volatiles in the final stages of imidization where the resin is extremely viscous, one would expect better results if the final step in curing were an addition (no by-products) rather than a condensation reaction. This addition step can be done if the imidization reaction is completed at an early stage in the curing process where the resin is still fusible and if addition-curing functionality is incorporated into the backbone polymer to be used for a final cross-linking reaction. The matrix resins which have been developed from this approach are referred to as *addition polyimides*. They resemble conventional condensation polyimides in having polyimide backbone structures and differ in having relatively low molecular weight oligomeric ("prepolymer") backbone chains end-capped with a group capable of undergoing an addition reaction during cure to yield a cross-linked resin.

The first commercial examples of addition polyimides were the bismaleimides introduced by Rhône-Poulenc and the P13N polyimides, developed by TRW/NASA [113] and introduced by Ciba-Geigy. Today the bismaleimides are probably the most widely used commercial polyimides for composite applications [114]. The P13N polyimide, on the other hand, has disappeared from the commercial market, but its basic structure, shown in Figure 3.3-38, is used in the form of PMR-15 polyimide, to be discussed later.

The chemistry of bismaleimides is outlined in Figure 3.3-39; the basic polyimide structure is the bismaleimide from the reaction of maleic anhydride with methylene dianiline. Although this bismaleimide will homopolymerize thermally through the maleic unsaturation [115], the resulting cross-linked structure is a very brittle matrix resin. However, if an appropriate amount of an aromatic diamine is copolymerized with the bismaleimide by Michael addition at the double bond, one can obtain a controlled degree of chain extension to form oligomers capped by maleimide groups, through which the system can be thermally cross-linked

PMDA
PYROMELLITIC DIANHYDRIDE

DPE
4,4'-DIAMINODIPHENYL ETHER

BTDA
BENZOPHENONE
TETRACARBOXYLIC DIANHYDRIDE

DDM
4,4'-DIAMINODIPHENYL METHANE

BTDE
BENZOPHENONE TETRACARBOXYLIC
ACID DIETHYL ESTER

PPD
p-PHENYLENE DIAMINE

DABP
3,3'-DIAMINOBENZOPHENONE

NE
MONOMETHYL ESTER OF
5-NORBORNENE-2,3- DICARBOXYLIC ACID

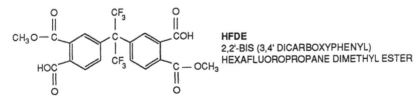

JEFFAMINE AP-22

HFDE
2,2'-BIS (3,4' DICARBOXYPHENYL)
HEXAFLUOROPROPANE DIMETHYL ESTER

FIGURE 3.3-37. Ingredients used in polyimides.

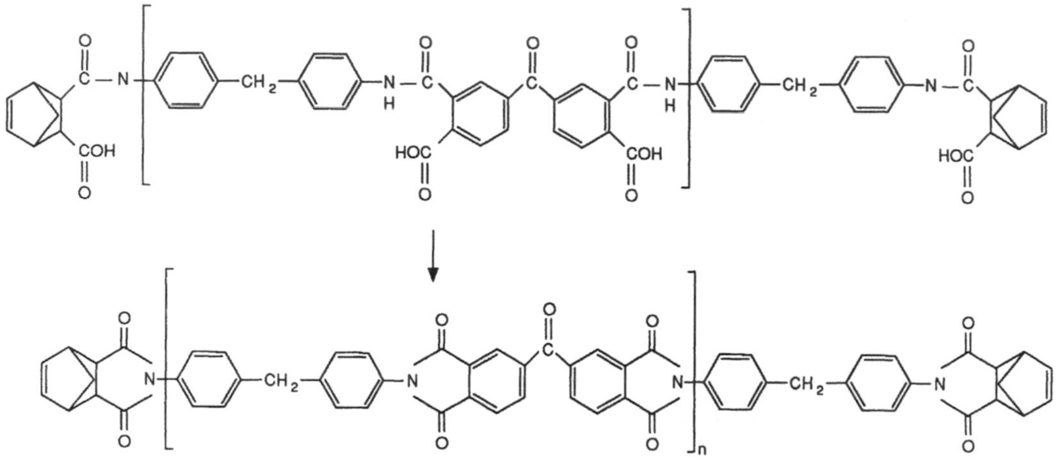

FIGURE 3.3-38. P13N chemistry.

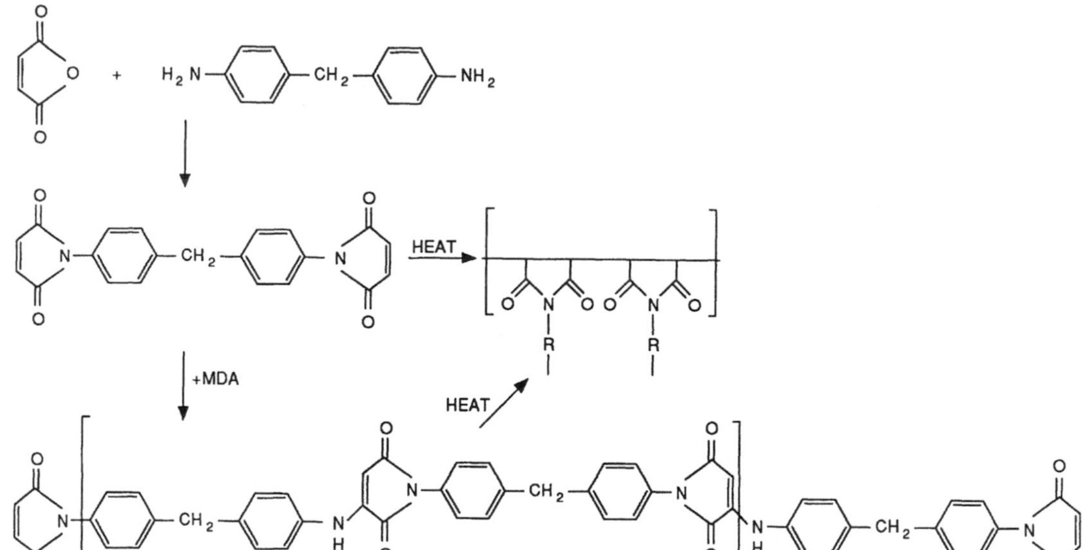

FIGURE 3.3-39. Bismaleimide chemistry.

to a less brittle, thermally stable matrix resin. The commercial polymers based on this chemistry (e.g., Kerimid 601) are mixtures of the bismaleimide with an aromatic amine such as methylene dianiline dissolved in a high boiling point, polar solvent (N-methylpyrollidone, NMP) for prepregging. Since the bismaleimide is fully imidized as received, all subsequent reactions are additions, and no by-products are evolved in staging or curing; the major problem in achieving low void, thermally stable resins is removing the polar solvent [116].

The final stages of cure involve an addition reaction through the maleimide unsaturation, as shown in Figure 3.3-39. This reaction occurs at temperatures above 175°C and leads to the formulation of the cross-linked resin.

PMR-15

A second method of obtaining an addition polyimide is to use dianhydrides and diamines similar to those used in condensation polyimides but to limit the molecular weight of the polyamic acid by capping with a monofunctional anhydride which also serves as a site for subsequent cross-linking. This general approach is the basis for most of the non-bismaleimide addition polyimides being used today; the most important examples are PMR-15 (and modified versions) and LARC-160.

The route to PMR-15 is outlined in Figure 3.3-40; note that the final structure is the same as that for P13N (Figure 3.3-38). The difference is in the route used to prepare the resin; in P13N, the imidization is completed and the capped, low molecular weight polymer is dissolved in a solvent for prepregging—subsequent solvent is removed, and curing occurs through the terminal unsaturated groups. In PMR-15, the impregnating solution is a mixture of the monomers, NE, MDA, BTDE (see Figure 3.3-37), with the correct stoichiometric ratio to give a polymer with the desired molecular weight (e.g., $MW = 1500$ for PMR-15); as shown in Figure 3.3-40, this monomer mixture is then converted via the polyamic acid stage to the fully imidized stage prior to final addition-type cross-linking. Since the imidization is completed while the resin is fusible and since the solvent is methanol, removal of volatiles is relatively easy, and low void content laminates can be prepared. The name "PMR" refers to Polymerization of Monomeric Reactants [117].

Processing of PMR-15 is accomplished by conventional vacuum bag autoclave or compression molding. Cure temperatures are usually less than 316°C (600°F), and cure times are 1–3 hours. Curing studies [118] have shown the presence of a melt-flow region in the early stages of cure that has been correlated (by DSC) with endothermic transitions due primarily to melting of monomers and initially formed oligomers. These studies also showed the exotherm due to the addition cross-linking reaction to be centered at about 340°C. This is presumably a reverse Diels-Alder reaction generating cyclopentadiene and a maleimide end group, which then copolymerize with the unsaturation in the other end groups to give an alicyclic copolymer cross-link [119].

LARC-160

A modified version of an addition polyimide based on the PMR approach has been introduced by NASA/Langley and designated LARC-160 [120]. The chemistry of this resin system is essentially the same as outlined in Figure 3.3-40 for PMR-15. The major differences are that LARC-160 uses a different aromatic diamine from that used in PMR-15 (i.e., Jeffamine AP-22 rather than methylene dianiline) and, in addition, uses the ethyl rather than methyl esters of the anhydrides. Both solvent and hot-melt prepregging have been demonstrated for LARC-160, and prepregs are reported to be more epoxy-like than are those of PMR-15 [120]. The chemistry of the imidization of LARC-160 has been studied by Delos et al. [121], who found that both esters (NE and BTDE) formed the imide directly, and no amide was present in quantity at any stage of reaction. The bisnadimide was a major product of the polymerization under all conditions. Aging of the LARC-160 monomer mixture was shown to affect the cure of the resin. Aging effects on cured resin properties have also been reported [122].

The processing and properties of LARC-160/graphite were studied by Mace et al. [123], who described tooling and manufacturing techniques. A cure cycle from this study is shown in Figure 3.3-41.

Other Addition Polyimides

PMR-15 and LARC-160 are the most widely used of the addition polyimides utilizing PMR chemistry, and most of the detailed polyimide composite evaluations have used one or the other of them as matrix resins.

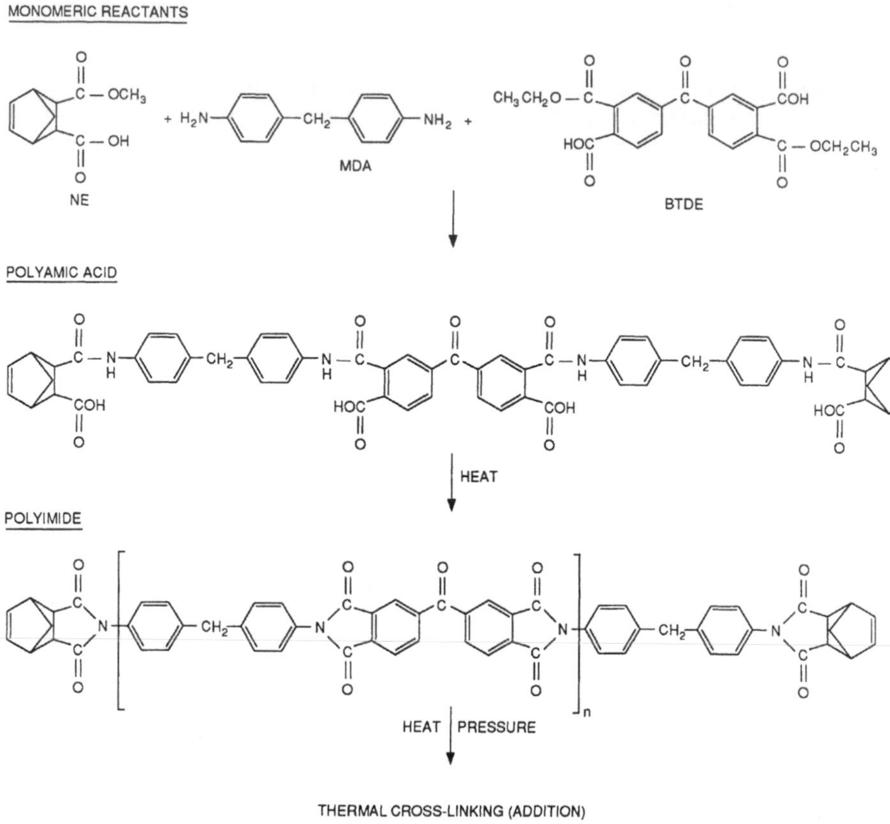

MONOMERIC REACTANTS

POLYAMIC ACID

HEAT

POLYIMIDE

HEAT | PRESSURE

THERMAL CROSS-LINKING (ADDITION)

FIGURE 3.3-40. PMR-15 chemistry.

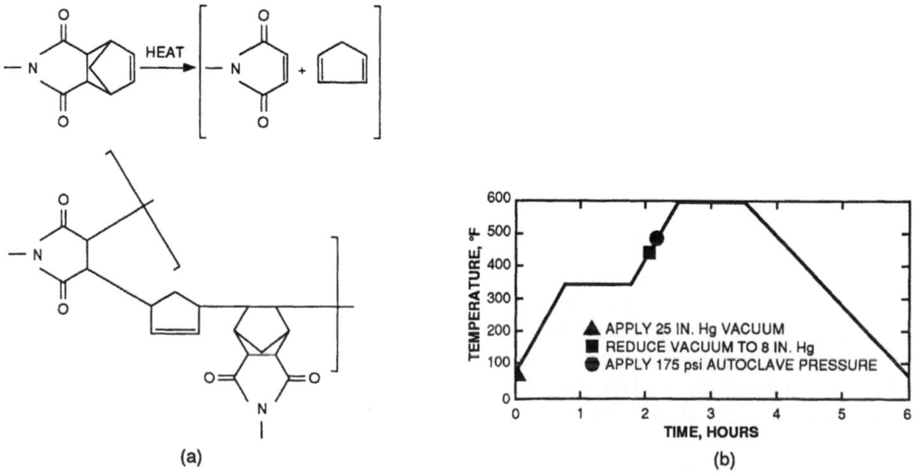

(a)

▲ APPLY 25 IN. Hg VACUUM
■ REDUCE VACUUM TO 8 IN. Hg
● APPLY 175 psi AUTOCLAVE PRESSURE

(b)

FIGURE 3.3-41. Autoclave cure cycle for LARC-160 from reference [123].

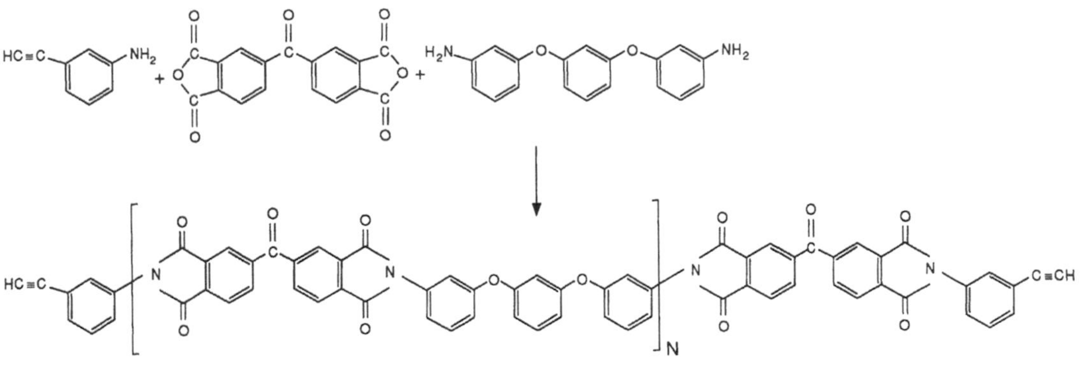

THERMID 600
(N = 1)

FIGURE 3.3-42. Acetylene-terminated polyimide chemistry.

However, a number of other polymers have been formulated, usually to achieve some specific improvement in properties or processing. Foremost among these polymers is PMR-II, which has been referred to as a "second generation" PMR polyimide [117]. The route to PMR-II is generally the same as that to PMR-15 (see Figure 3.3-40) except that two of the three monomers are replaced by different structures; the anhydride is HFDE instead of BTDE, and the diamine is PPD instead of MDA (names and structures shown in Figure 3.3-37). PMR II is reported to have a higher T_g and better property retention after aging at 600°F than PMR-15 [117].

Acetylene-Capped Polyimides

These polymers are another example of the general principle of capping a relatively low molecular weight oligomer with an end group capable of undergoing an addition-type cross-linking reaction. In this case, the end group is acetylenic and the backbone polyimide is highly aromatic; the preparation shown in Figure 3.3-42 is for the structure which was commercially available as Thermid 600. Various modifications of this structure have also been evaluated [124]. They include changes in the value of *"N"*, changes in the dianhydride structure, and changes in the structure of the acetylenic end cap group [125].

These capped polyimides can be used for prepregging in either the fully imidized form (from NMP or DMF) or in the polyamic acid form (from acetone). Use of the polyamic acid for prepregging gives a system amenable to vacuum bag and autoclave processing. Imidization occurs in the 100–170°C range prior to final cure. Strength properties of laminates prepared by this route are somewhat lower than those for laminates prepared from the fully imidized prepolymer [124]. Successful hot-melt prepregging has also been reported [126] to give low volatile content prepregs.

Curing of acetylene-terminated polyimides occurs at temperatures above 250°C and is thought to proceed by cyclization of acetylene groups to give an aromatic ring. This cure mechanism differs from those discussed previously in that the cross-link is an aromatic structure offering more thermal stability than the aliphatic cross-links in the PMR or bismaleimide polyimides. TGA weight loss measurements on the neat resin suggest that the acetylene end cap gives cured resin thermal stability similar to the all-aromatic condensation polyimides. However, other studies on the cross-linking efficiency of various end caps for addition-curing polyimides have suggested that vinyl groups are as effective as ethynyl groups in providing thermally stable cross-links [127].

Cure cycles for the acetylene-terminated polyimides utilize high initial heat-up rates to insure exceeding the melting point of the resin prior to the cross-linking reaction, which can occur in the solid state [126]. A cure temperature of 250°C and postcuring at temperatures up to 370°C give T_g values as high as 390°C.

Thermoplastic Polyimides

Although the condensation polyimides are thermoplastic by virtue of being linear, uncross-linked poly-

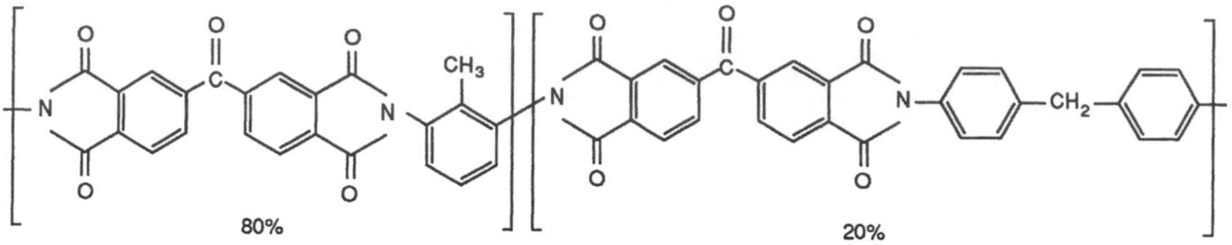

UPJOHN POLYIMIDE
2080

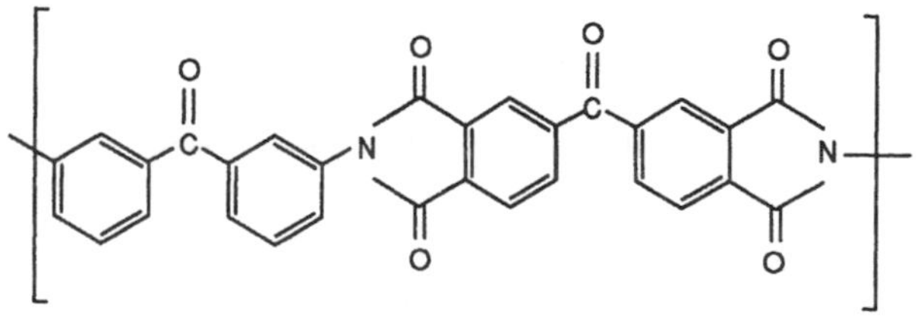

LARC-TPI

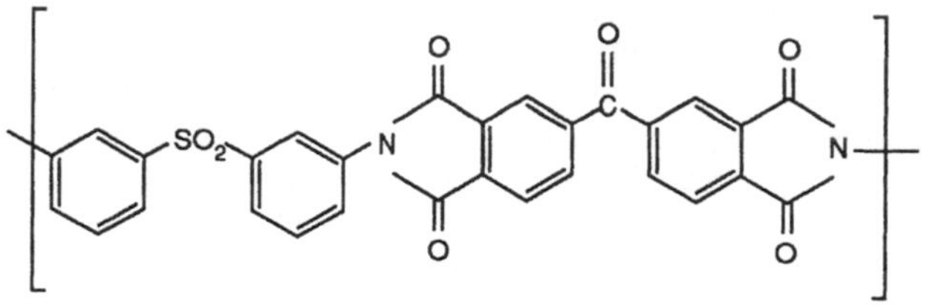

POLYIMIDESULFONE

FIGURE 3.3-43. Thermoplastic polyimides.

mers (those based on BTDA may react to some extent through the carbonyl functionality), they are not melt processable. One route to melt processability is to decrease chain rigidity, which can be done by incorporating thermally stable, flexibilizing groups into the polymer backbone by using multi-ring dianhydrides and diamines connected by those groups or by using isomeric mixtures of diamines to depress softening temperatures. An early example of this approach was NR-150 (DuPont), Figure 3.3-40, which has a T_g of about 280°C and can be melt processed [128]; these

polyimides are no longer commercially available, but the approach has been used for other structures.

The structures of several current thermoplastic polyimides are shown in Figure 3.3-43. Upjohn's Polyimide 2080 is a block copolymer of BTDA and the diisocyanates form a 20/80 mixture of methylene dianiline and toluene dianilines. This material is supplied as a fully imidized solution which can be used for prepregging. Laminate formation involves removal of solvent and consolidation of the lay-up; no chemical reactions occur. Cycle times are relatively short, and low void, high quality laminates have been reported. Air aging of Polyimide 2080/glass laminates shows good thermal stability at high temperatures: at 375°C (707°F), the laminate retained 50% of its initial flexural strength for 50 hours [129].

Another thermoplastic polyimide currently being evaluated, designated LARC-TPI, is shown in Figure 3.3-43. Its structure utilizes both multi-ring diamine and dianhydride, plus meta-orientation in the diamine, to disrupt the chain symmetry—the monomer structures are shown in Figure 3.3-37. LARC-TPI has a T_g of 266°C and good thermal stability in air aging at 300°C (2–3% weight loss after 550 hours). Although it is being evaluated primarily as an adhesive for structural applications, LARC-TPI has been investigated as a matrix resin with graphite fibers [130].

Many other variations are possible by changing the diamine portion of the polymers, and one of these is the polyimidesulfone shown in Figure 3.3-43. Its structure differs from LARC-TPI only in having sulfone replace carbonyl as the connecting link between the two rings of the diamine. It has been reported to have moderately good toughness (G_{1c} = 1.4 kJ/m²) and thermal stability [130] but has not been thoroughly evaluated in laminates.

The most recent thermoplastic polyimides to be introduced are known as K-polymers [128,131]. Two versions are available, K-1 and K-2, but only K-1 is discussed here. K-1 polyimide is an amorphous polymer with a T_g of 210°C; neat resin properties are shown in Table 3.3-41. In addition to good mechanical properties, the polymer has high toughness (G_{1c} = 6.3 kJ/m²), low moisture absorption, and good retention of properties under hot/wet conditions. These resin properties result in good laminate performance in matrix-dominated tests; for example, flexural and short-beam shear-strength retentions for K-1/graphite were 55–56% at 350°F in the moisture-saturated condition. The high toughness of the resin resulted in good laminate compression-after-impact performance even at high levels of impact energy. Although K-1 polyimide is an amorphous thermoplastic, it shows good resistance to a variety of chemicals, including those commonly encountered in aircraft applications.

Processing of K-1 polyimide by both vacuum bag/press molding of a tacky, drapable prepreg and automatic lay-down of a fully consolidated tape has been reported [128,131]. Processing temperatures are in the 320–370°C range. K-2 polyimide is a higher T_g resin

(~270°C) that also has higher toughness (G_{1c} = ~14 kJ/m²) than K-1 but that is at an earlier stage of development.

Properties

Mechanical. Properties of neat resins are summarized in Table 3.3-42. In general, the polyimides are not as well characterized in neat resin form as are the epoxies, partly due to the difficulty of preparing good specimens. Room temperature tensile properties for the polyimides in Table 3.3-42 are in the same range as those of the epoxies typically used in high-performance applications: tensile strengths range from 7 to 14 ksi, tensile moduli from 450–600 ksi, and tensile elongations from 1 to 3%. Flexural properties of bismaleimides, shown in Table 3.3-43, are maintained fairly well up to 250°C (60% retention of flex strength, 78% retention of modulus) and also down to −200°C.

Mechanical properties for T-300 bismaleimide laminates are shown in Table 3.3-44. Retention of both compressive and shear properties is 80–100% at 450°F. This resin (Hexcel F-178) is reported [132] to be a bismaleimide modified with triallylisocyanurate (TAIC). Similar data for Celion 6000/PMR-15 are shown in Table 3.3-45 for three lay-ups tested from −157°C to 316°C. The data for the (+45)₂ lay-up are plotted in

Table 3.3-41. Neat resin properties of K-1 polyimide (data from references [128,131]).

Tensile Strength, ksi	
23°C	15.1
93°C	11.7
Tensile Modulus, ksi	
23°C	345
93°C	321
Tensile Elongation, %	
23°C	6.4
93°C	5.1
T_g, °C	
Dry	210
Wet	191
G_{1c}, kJ/m²	6.3
Oxygen index	50
Char yield, %	60
Density, g/cm³	1.37
Moisture absorption, % (equil. in H₂O immersion)	0.87

Table 3.3-42. Properties of polyimide neat resins.

		PMR-15	PMR-II	Kerimid 601	Thermid 600	V-278A	LARC-TPI	Polyimide 2080
Tensile Modulus	10^3 psi	470	600		550			–
Tensile Strength	10^3 psi	8.10	11.2	7–8.5	14	11.3		17.1
Tensile Elong.	%		2.5	<1	2.6	6.6		10
Compressive Modulus	10^3 psi	–						295
Compressive Strength	10^3 psi	27.2		29	to 66.0			30
Flex Strength	10^3 psi			14	18–21		23	29
Flex Modulus	10^3 psi			455	650			481
CTE	PPM/°C	28		49 0–300°C				50
Specific Gravity		1.30	1.42	1.30		1.27	1.37–1.40	1.4
Dielectric Constant/ Freq. (H_3)				3.5/50 3.5/10^7				3.43 (60) 3.42 (10^3)
Dissipation Factor/ Freq. (H_3)				0.025/50 0.02/10^7				0.005 (60) 0.0018 (10^3)
T_g	°C	350(TMA)			285–340 (TMA)	370	266°C	310°C
Oxygen Index				35				44

Figure 3.3-44 to show retentions at low and high temperatures compared to room temperature values. At 316°C, the tensile strength and modulus retentions are 76% and 46%, respectively. Mace et al. [123] studied processing and properties of graphite/LARC-160 fabric laminates and reported that 90–93% of room temperature compressive and shear values were retained at 600°F; however, flexural strength retention was significantly lower, as shown in Table 3.3-46. Jones [134]

Table 3.3-43. Flexural properties of bismaleimide neat resins vs. temperatures (Kerimid 601; data from Rhône-Poulenc).

Flexural Modulus, ksi		% of RT Value
250°C	455	–
200°C	385	85
250°C	355	78
−100°C	540	119
−200°C	870	191
Flexural Strength, ksi		
25	14.2	–
200	10.0	70
250	8.5	60
−100	14.5	102
−200	14.3	101

examined a variety of graphite/LARC-160 lay-ups over a wide range of temperatures. The compression and tensile strength retentions of the ±45° laminates from this study are plotted in Figure 3.3-45 to show changes at both high and low temperatures in tests which are considered to be dominated by the resin shear strength. The wide discrepancy between the retentions in tension and compression was apparently due to a resin curing effect, since it disappeared after 125 hours of aging at 600°F (at which point the tensile and compressive strength retentions were 77% and 80%, respectively). The aging performance of these laminates is discussed further in a subsequent part. Delvigs et al. [133] compared PMR-15 and PMR-II, each with two different graphite-fiber types, from room temperature to 600°F (Table 3.3-47). Differences between the two resins were generally small, regardless of fiber type, but in all cases the retentions at 600°F were slightly higher for PMR-15.

A comparison of the delamination and environmental resistance of a commercially available, modified graphite/bismaleimide laminate and a state-of-the-art 350°F-curing structural graphite/epoxy laminate was made by Wilkins [135]. The polyimide was V-378A (U.S. Polymeric) and the epoxy was 3501-6 (Hercules). The major

Table 3.3-44. Properties of a Hexcel F-178/T-300, graphite/bismaleimide laminate at room temperature and 450°F [132].

		Test Direction	Temperature		% of RT Value
			RT	450°F	
Tensile					
Strength,	10^3 psi	0	66.5	71.7	108
Modulus,	10^6 psi	0	7.85	7.78	99
Compressive					
Strength,	10^3 psi	0	67.2	–	–
Strength,	10^3 psi	90	47.7	40.0	86
Modulus,	10^6 psi	0	7.98	–	–
Modulus,	10^6 psi	90	6.29	6.63	105
Flexural					
Strength,	10^3 psi	0	156.9	–	–
Modulus,	10^6 psi	0	11.3	–	–
Rail Shear					
Strength,	10^3 psi	90/0	28.0	22.3	80
Modulus,	10^6 psi	90/0	2.30	2.38	103
Interlaminar Shear Strength,	10^3 psi	0	5.99	–	–

Notes

Fiber Volume, 0/0: 61
Specific Gravity: 1.57
Molded Ply Thickness: 0.0053 in.
Lay-up: $[0(0/\pm45/90)_2,0]$

Table 3.3-46. Properties of LARC-160/T-300 fabric laminates (data from reference [123]).

Flexural Strength	ksi	MPa	% of RT Value
RT	71.2	491	–
177°C (350°F)	71.7	495	100
260°C (500°F)	71.7	495	100
315°C (600°F)	45.6	314	64
Compressive Strength			
RT	70.7	487	–
177°C (350°F)	72.3	498	102
260°C (500°F)	70.4	485	100
315°C (600°F)	65.8	454	93
Short Beam Shear Strength			
RT	4.8	33.1	–
177°C (350°F)	5.5	38.0	115
260°C (500°F)	5.7	39.1	119
315°C (600°F)	4.3	29.7	90

Material Specifications.

Fabric: 8 HS (24 × 25, 3K)
Fiber Volume: 56 ± 1%
Void Content: 2–5% (By Acid Digestion).
Density: 1.52–1.54 g/cm³.
Post Cure: 2 Hours/316°C (T_g = 348°C).

Table 3.3-45. Variation of Celion 6000/PMR-15 properties with temperature [133].

	Temperature, °C		
	−157	24	316
$[\pm45]_{2s}$			
Tensile strength, ksi	19.3	17.6	13.4
Tensile modulus, 10^6 psi	3.23	2.68	1.23
Poisson's ratio	0.686	0.776	0.813
G_{12}, 10^6 psi	0.959	0.678	0.304
$[90]_8$			
Tensile strength, ksi	4.61	4.02	2.57
Tensile modulus, 10^6 psi	1.52	1.24	0.860
$[\pm45]_{4s}$			
Tensile strength, ksi	21.0	19.1	13.8
Tensile modulus, 10^6 psi	3.23	2.49	–
Poisson's ratio	0.737	0.771	0.885

Table 3.3-47. Mechanical properties of graphite/PMR composites (unidirectional, 55% fiber volume) [133].

Interlaminar Shear Str.	Celion 6000		HTS-2	
	PMR-15	PMR-11	PMR-15	PMR-11
RT ksi	15.2	15.3	14.1	13.0
600°F ksi	7.2	6.8	7.6	6.0
% retention	46	44	54	46
Flexural Strength, ksi				
RT	187	191	233	241
600°F	104	105	146	128
% retention	56	55	63	53
Flexural Modulus, 10^6 psi				
RT	14.1	15.5	16.3	16.1
600°F	12.7	13.7	16.0	15.3
% retention	90	88	98	95

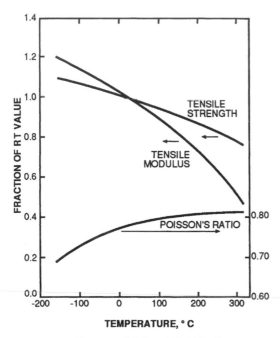

FIGURE 3.3-44. Variation of Celion 6000/PMR-15 properties with temperature ([±45]₂, laminates) [133].

conclusion from this study was that the polyimide gave about a 100°F advantage in environmental resistance over the epoxy. Tests included Mode I and Mode II toughness, wet and dry T_g, thermal spiking, shear properties, and hot/wet compression; the only test in which the polyimide was deficient was the Mode II toughness test, where the epoxy value was about 50% higher.

Odom and Adams [136] compared neat resins of bismaleimide and a structural epoxy (Hercules 3501-6) in tension-tension fatigue ($R = 0.1$) and estimated the maximum stress levels at room temperature for a life of 10⁶ cycles to be about 6 ksi for the polyimide vs. 5 ksi for the epoxy—both dry; at 88°C, the stress levels were about 4.3 ksi for the polyimide vs. 3.6 ksi for the epoxy. Torsional fatigue gave the same ranking of the resins (with all stress levels in the 4-6 ksi range). The static strength, the modulus, and the cure cycles used to prepare the specimens are shown in Table 3.3-48.

Fracture toughness values of several polyimide resins are shown in Table 3.3-49, arranged in order of increasing toughness relative to a 350° cure structural epoxy; it is observed that the least tough polyimide in this series is about 1.6 times tougher than the epoxy.

Thermal. Since one of the most common reasons for choosing a polyimide matrix resin is to obtain satisfac-

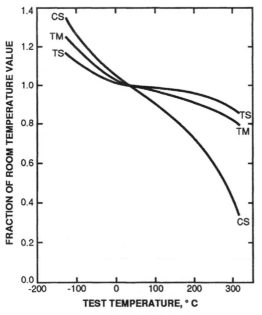

FIGURE 3.3-45. Properties of graphite/LARC-160 vs. temperature (±45)ₛ (Celion 3000, V$_f$ = 60%) [134].

Table 3.3-48. Comparison of tensile and shear properties of bismaleimide and epoxy neat resins [136].

	Epoxy (Hercules 3501-6)	Bismaleimide (Hercules X-4001)
Tensile Strength, ksi		
RT	8.6	9.6
88°C	7.6	8.3
% retention	88	86
Tensile Modulus, ksi		
RT	500	520
88°C	520	400
% retention	104	77
Shear Stress (Torsion), ksi		
RT	11.4	11.0
88°C	12.1	9.9
% retention	106	90
Shear Modulus (Torsion), ksi		
RT	260	260
88°C	250	200
% retention	96	77
Cure Cycles		

3501-6 epoxy: 50 min/107°C + 5 hr/140°C + 3 hr/177°C
X-4001 bismaleimide: 50 min/107°C + 5 hr/140°C + 4 hr/177°C + 8 hr/204°C

tory performance at elevated temperatures, much evaluation work has involved establishing property retentions at elevated temperature for both short- and long-term exposures. The structural features in the resin which provide good short-term retention of properties at elevated temperatures are not necessarily the same as the features which make the resin resistant to long-term thermal and/or oxidative effects that determine its lifetime in service. In general, studies on polyimides have established maximum "use" temperatures of about 480°F for bismaleimides and about 600°F for PMR and condensation types. Most evaluations of polyimide performance involve determination of a useful lifetime at elevated temperatures; therefore, essentially all data reported in this discussion are for tests from room temperature (or below) to a maximum of 600°F. Of course, if the interest is in lifetimes of minutes instead of weeks or months, one can trade longer times for higher temperatures; for example, some investigators [138] consider that satisfactory structural performance may be obtained at 800–900°F for about one minute vs. 125 hours at 600°F or 70,000 hours at 350–450°F with present state-of-the-art polyimides. Most of the evaluations covered in this discussion show that, at the upper end of the temperature ranges studied with a given resin, the best that can be expected is lifetimes of a few hundred hours, depending on what degree of property retention is considered for a useful life.

In all aging studies, the inherent thermal stability of the neat resin is obviously of fundamental importance: one useful method of comparing resins is by TGA (thermogravimetric analysis), which is essentially an instrumental method of following weight changes vs. time in a controlled atmosphere and at a controlled heating rate [138]. TGA analyses in an inert atmosphere (e.g., nitrogen) give an indication of the thermal stability of the resin; the same analysis in air is a measure of thermal stability *plus* susceptibility of the resin to oxidative attack. Table 3.3-50 lists the TGA temperatures at which various polyimide resins suffered 10% weight loss in either air or nitrogen atmosphere; in addition, the table shows the temperature at which the *rate* of weight loss peaked. The order of resin stability by this test is in good agreement with results of long-term laminate aging studies. It shows the condensation polyimides to be the most stable and the bismaleimides to be the least thermally stable. Other studies [140] agree with these results and, in addition, show that the overall activation energies for degradation are highest for the

Table 3.3-49. Fracture toughness of polyimides vs. a structural epoxy [137].

Resin Type, Generic	Trade Designation	Relative* G_{1c}
Epoxy, 350°F	NARMCO 5208	1.0
Bisaleimide	Hexcell F-178	1.6
Acetylenic Polyimide	Thermid 600	2.5
PMR Polyimide	PR-15	2.6
Thermoplastic Polyimide	Upjohn 2080	11
Thermoplastic Polyimide	NR-150 B2	28
Polyamide-imide	Torlon 4000	45

*G_{1c} for NARMCO 5208 is 0.076 kJ/m².

condensation types, intermediate for the PMR types, and lowest for the bismaleimide types.

Comparisons of PMR-15 and PMR-II showed that weight loss of PMR-15 neat resins was approximately 14% after 1,000 hours at 600°F in air; PMR-II, at three values of "n" (1.30, 1.67, 2.50) had generally higher weight loss than PMR-15 up to 1,000 hours except for $n = 2.50$, which had lower weight loss than PMR-15 after the first 700 hours [117].

Aging effects on a glass/bismaleimide laminate are shown in Figure 3.3-46; after 2,650 hours at 250°C, the laminate retained 50% of initial flex strength (both initial and final values measured at room temperature) and suffered a weight loss of about 7%. Data for a commercially available modified graphite/bismaleimide laminate is shown in Table 3.3-51; at 232°C (450°F), the laminate retains 88% of its initial flex strength and 100% of initial shear strength for nine months, while

Table 3.3-50. Polyimide neat resin weight loss by thermogravimetric analysis (TGA) [139].

Resin	Temp. for 10% wt. Loss, °C		Temp. for Maximum Rate of Weight Loss in Air, °C
	Air	N₂	
Skybond 700	530	550	570
Kapton Film	520	560	570
Polyimide 2080	490	535	550
Thermid 600	460	525	490
PMR-15	420	450	515
Kerimid 353	415	450	525
Kerimid 601	400	400	550

Heating Rate: 2°C/minute.

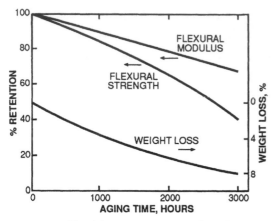

FIGURE 3.3-46. Flexural property retention and weight loss vs. aging time at 250°C. For Kerimad 601 laminates (S-181 fabric) (data from Rhône-Poulenc).

suffering only 3.4% weight loss. This resin (V-378A from U.S. Polymeric) is an example of a bismaleimide useful at temperatures higher than provided by 350°F epoxies but not as high as provided by other polyimides.

An extensive evaluation of LARC-160/graphite as a replacement for aluminum in various Space Shuttle

Table 3.3-51. Aging performance of a modified graphite/ bismaleimide laminate (USA Polymeric V-378A polyimide/ T-300; $V_f = 64–66\%$) (data from reference [141]).

	Test Temperature, °C		
	RT	177	232
0° Flexural Strength, 10³ psi			
Initial	298	245	233
Aged 6 months at 177°C	320	271	–
% of initial value	107	111	–
Aged 6 months at 232°C	275	–	180
% of initial value	92	–	77
Aged 9 months at 232°C	267	–	205
% of initial value	90	–	88
Short Beam Shear Strength, 10³ psi			
Initial	15.8	10.1	7.7
Aged 9 months at 232°C	15.5	–	8.6
% of initial value	98	–	112
Weight Loss, %			
Aged 6 months at 177°C	0.57		
Aged 6 months at 232°C	2.3		
Aged 9 months at 232°C	3.4		

parts was carried out by Jones [134]; in addition to processing and property studies, aging tests were performed at 316°C (600°F), and properties of aged laminates were measured at four test temperatures, as shown in Table 3.3-52. For unidirectional laminates, 0° compressive strength after aging (125 hours at 600°F) was 89% of the unaged value, both measured at 600°F; the corresponding 90° tensile strength retention was 80%. For ±45° laminates, the compressive strength at 600°F was higher after aging than before (by >2X) which was attributed to further postcuring of the resin. However, the tensile strength of the same ±45° laminate showed no such postcuring effect; its strength retention after aging was 74% of the unaged value, both measured at 600°F.

Aging results for graphite/PMR-15 are shown in Table 3.3-53. The fiber type in these tests was Celion 6000 and the fiber volume fraction was 60% [142]. The laminates maintained at least 50% of RT flexural and shear properties for 1,500 hours at 600°F; in addition, a PMR composition modified for improved flow was found to be as thermo-oxidatively stable as PMR-15.

The effects of 50,000 hours of elevated temperature aging on graphite/epoxy and graphite/polyimide were determined by Kerr and Haskins [143] for laminates with AS graphite 3501-5/epoxy and HT-S graphite/Hercules 710 polyimide. The results from this study are highlighted below.

Graphite/Epoxy

250°F   No effects for 10,000 hours. Matrix degradation began between 10,000 and 25,000 hours and was severe after 50,000 hours.

350°F   Matrix degradation began between 1,000 and 5,000 hours. Matrix crumbled away after 5,000 hours. Reduction of pressure from 1.47 to 2 psi delayed but did not eliminate matrix degradation.

Graphite/Polyimide

450°F   No effects for 25,000 hours; decrease in tensile strength observed after 50,000 hours but no observed matrix degradation.

550°F   Tensile strength decreased after 10,000 hours; partial delamination and high weight loss after 25,000 hours.

Table 3.3-52. Effect of temperature and aging time on properties of Celion 3000/LARC-160 polyimide laminates ($V_f = 60\%$) [134].

Unidirectional Laminate		Test Temperature, °F							
		Postcured 4 hr/600°F				Aged 125 hrs at 600°F			
		−270	RT	400	600	−270	RT	400	600
0° Compressive									
Strength	10³ psi	229	174	148	112	227	209	105	99.6
Modulus	10⁶ psi	20.1	18.3	17.8	18.0	20.7	18.6	22.7	21.5
Ult. Strain	%	1.42	1.05	0.96	0.64	1.28	1.29	0.47	0.47
90° Tensile									
Strength	10³ psi	4.75	3.07	2.12	2.42	6.53	4.85	1.66	1.93
Modulus	10⁶ psi	1.47	1.47	1.09	0.798	–	–	–	–
Ult. Strain	%	0.33	0.33	0.20	0.37	–	–	–	–
(+45°) Laminate Tensile									
Strength	10³ psi	27.1	23.7	22.1	20.2	23.0	19.5	19.3	15.0
Modulus	10⁶ psi	3.74	3.00	2.73	2.39	–	–	–	–
Ult. Strain	%	0.74	–	–	–	–	–	–	–
Compressive									
Strength	10³ psi	33.7	25.0	17.9	8.7	30.5	23.0	20.2	18.4
Modulus	10⁶ psi	2.99	2.36	1.60	1.1	3.05	2.52	2.07	1.87
Ult. Strain	%	1.50	3.00	4.17	2.27	1.60	1.47	–	–

The graphite/polyimide performance at 550°F was clearly superior to graphite/epoxy at 350°F in these tests. The lifetime of the polyimide composite at 550°F appears to exceed one year.

Environmental Resistance. Chemical resistance of polyimides is generally good for both condensation and addition types. Typical organic solvents which are encountered by aircraft parts have little or no effect on polyimides; indeed, mild acids and bases do not severely attack polyimides. However, the polyimide ring itself is susceptible to alkaline hydrolysis in the presence of strong bases even at room temperature. Dine-Hart et al. [144], for example, showed that a typical condensation polyimide slowly hydrolyzes in sodium hydroxide at room temperature to give the sodium salt of the original polyamic acid which, on recovery, can be reconverted to a polyimide of lower molecular weight. Complete hydrolysis to diamine and tetra-acid occurred at 80°C.

Table 3.3-53. Retention of graphite/PMR-15 properties after air aging at 600°F [142]
(fiber volume fraction: 60%; fiber type: Celion 6000).

	Test Temperature		% of Unaged RT Value	% of Unaged 600°F Value
	RT	600°F		
Flexural Strength, ksi				
Initial	231	134	58	100
After 1,000 hr. at 600°F	–	147	64	110
After 1,500 hr. at 600°F	–	116	50	87
Flexural Modulus, 10⁶ psi				
Initial	16.2	14.6	90	100
After 1,000 hr. at 600°F	–	12.3	76	84
After 1,500 hr. at 600°F	–	8.3	51	57
Interlaminar Shear Strength, ksi				
Initial	14.2	6.46	45	100
After 1,000 hr. at 600°F	–	6.54	45	101
After 1,500 hr. at 600°F	–	4.40	31	68

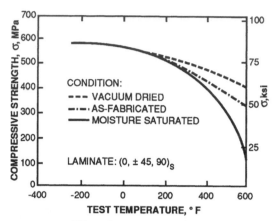

FIGURE 3.3-47. Effect of temperature and moisture on compressive strength of a Celion/PRM-15 laminate [139].

The effect of water immersion at 60 to 100°C on glass/condensation polyimide laminates was found [145] to be due both to reversible plasticization of the resin and to irreversible damage attributed to resin hydrolysis. The activation energy for the degradation was approximately the same as for hydrolysis of Kapton polyimide film (about 16 kcal/mole). Exposure in steam was found to be just as severe as immersion in boiling water.

The equilibrium moisture absorption of PMR-II (neat resin) was found to be 3.8% at 97% R.H./65°C for specimens that were not postcured [117]. Postcure at 600°F increased the equilibrium moisture content to 4.3%, presumably due to increased microcracking.

Moisture absorption data for bismaleimide and graphite/PMR-15 laminates are shown in Table 3.3-54. Diffusion coefficients vary with temperature and are relatively constant with varying relative humidity at constant temperature, as expected. The equilibrium moisture contents at 95% R.H. are about 32–38% higher than for epoxy laminates (with an MY-720-based epoxy matrix); in addition, the diffusion coefficients are higher than those of the epoxy laminates (by 8–10×) with approximately the same temperature dependence [146].

The effect of moisture and temperature on the compressive strength of a graphite/PMR-15 laminate is shown in Figure 3.3-47 for test temperatures from −250°F to 600°F. The moisture environment in these tests varied from dry to saturation at 100% R.H. At 600°F, the compressive strength retentions varied from 71% (dry) to 22% (saturated at 100% R.H.). Moisture in the laminate had no effect on compressive strength until the test temperature reached almost 200°F. At a test temperature of 300°F, even the completely saturated panel retained about 85% of its room temperature value, which is considerably better performance than a typical structural epoxy provides under similar conditions.

The hot/wet performance of graphite/PMR-15 and graphite/LARC-160 laminates is shown in Figure 3.3-48, where short beam shear strength of the moisture saturated laminates is plotted vs. temperature over the range 75–600°F. On a relative basis, the performance of PMR-15 is better (42% retention vs. 29% for LARC-160), but the LARC-160 maintains higher absolute shear strength from room temperature to about 300°F.

Electrical Properties. The polyimides have generally good electrical properties, including high dielectric strength, low dielectric constant, and low dissipation factor; they retain these properties over a wide temperature and frequency range. An example of the performance of a condensation polyimide laminate is shown in Figure 3.3-49, where both dielectric constant and dissipation factor variations are shown for temperatures

Table 3.3-54. Moisture absorption of bismaleimide and T-300 graphite/PRM-15 laminates [146] (fiber volume fraction: 62 ± 3%).

Exposure Conditions		PMR-15/T-300		F-178/T-300 (Bismaleimide)	
Temperature, °F	RH, %	Equil. Moisture Content, %	Diffusion Coefficient $D, 10^{-8}$ cm²/sec	Equil. Moisture Content, %	Diffusion Coefficient $D, 10^{-8}$ cm²/sec
77	95	1.65	0.8	1.72	0.5
125	95	1.58	1.9	1.68	1.3
160	95	1.46	4.4	1.68	3.3
160	75	1.22	3.9	1.36	3.4
160	51	0.84	3.7	0.88	3.3

from 25°C to 300°C at an X-band radar frequency (8.5 GHz); the dielectric constant is almost constant over this range, and the dissipation factor variations are quite low. The property combination of high resin glass transition temperature and low dielectric constant/dissipation factor is a major reason for polyimides being used in radomes [147] and printed circuit boards, two of the most important electrical applications for these resins. Good environmental resistance and low thermal expansion coefficients are other important attributes for these applications.

The use of bismaleimide resins in printed circuit board (PCB) technology has become more widespread where temperature requirements for the PCB have increased and standard FR-4 epoxy boards no longer suffice. The chemistry of the bismaleimide resins allows them to be used to modify the usual epoxy resins to achieve both higher T_g and lower CTE (coefficient of thermal expansion) without a sacrifice in electrical properties and chemical resistance [148–150].

The low resin CTE is of particular importance in reducing out-of-plane (Z-axis) thermal expansion (which is matrix-dominated) in boards which are designed to give specific X-Y expansions by controlling fiber volume. Excessive Z-axis expansion in multilayer printed circuit boards may lead to various problems, including loss of electrical continuity from one layer to another. The use of polyimides allows Z-axis CTE values in the 45–60 PPM/°C range to be achieved in boards which are useful to 230°C vs. 130°C for epoxy-based boards.

Some electrical properties of a bismaleimide neat resin are shown in Table 3.3-55 both for specimens aged at 480°F for 2,000 hours and for unaged controls. The dielectric constant at both 50 Hz and 10 MHz is unchanged after aging, and the dissipation factor

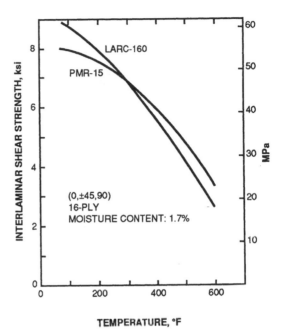

FIGURE 3.3-48. Effect of temperature on interlaminar shear strength of moisture-saturated graphite/polyimide laminates (PMR-15, LARC-160, and Celion graphite fibers).

Table 3.3-55. Neat resin electrical properties of a bismaleimide resin (Kerimid 601 data from Rhône-Poulenc).

Dielectric Constant	Initial	After 2,000 hrs at 480°F
50 Hz	3.5	3.5
10 MHz	3.5	3.5
Dissipation Factor		
50 Hz	0.025	0.04
10 MHz	0.02	0.01
Volume Resistivity, ohm-cm	1.1×10^{16}	6×10^{13}
Dielectric Strength, volts/mil	750	–

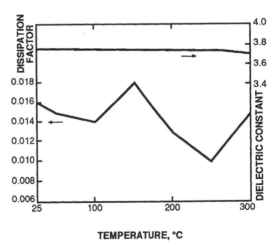

FIGURE 3.3-49. Dielectric properties vs. temperature for a condensation E-glass/polyimide laminate (skybond 700; 8.5 GHz; data from Monsanto).

Table 3.3-56. Representative commercial polyimide prepregs.

Supplier	Designation	Type	Upper Service Temp. °F	Comments
American Cyanamid	CYCOM 3002	Condensation	700	Best head resistance
American Cyanamid	CYCOM 3004	Condensation	<700	Better processability than 3002
American Cyanamid	CYCOM 3005	Addition; LARC-160	600	
American Cyanamid	CYCOM 3006	Addition; PMR-15	600	
American Cyanamid	CYCOM 3007	Bismaleimide	450	Epoxy-like processing
American Cyanamid	CYCOM 3102	Bismaleimide	–	Epoxy-like processing
American Cyanamid	CYCOM X-3100	Bismaleimide	–	Epoxy-like processing
Ferro	CPI-2214	Bismaleimide	500	Electrical applications
Ferro	CPI-2237	Addition; PRM-15	600	Light tack; radomes, compressor blades
Ferro	CPI-2249	Condensation	550	Low pressure cure; radomes, gen. purpose
Ferro	CPI-2255AF	Addition; PMR-15	600	Supported adhesive film
Ferro	CPI-2272X	Bismaleimide	450	Microcrack, moisture resistant; aircraft applications
Ferro	CPI-2274	Addition; LARC-160	600	Good tack and drape; radomes, structural
Hexcel	F-170	Condensation	600	
Hexcel	F-173	Condensation	600	
Hexcel	F-174	Condensation	600	
Hexcel	F-178	Bismaleimide	475	
Fiberite	966	Addition; PMR-15	600	
Fiberite	978	Addition; LARC-160	600	
Fiberite	987	Bismaleimide	450	
US Polymeric	V-378A	Modified Bismaleimide	450	Epoxy-like processing
Narmco	5245C	Modified Bismaleimide	350	Epoxy-like processing

changes are relatively small. The resin in these tests, Kerimid 601, is the base material used in many prepreg formulations for both structural and circuit board applications.

Flammability and Smoke Properties. Polyimides, together with phenolics, are the best readily available matrix resins for achieving good flammability and smoke performance. Oxygen index values for bismaleimide and PMR types are in the 30–35 range for the neat resins, and both have high char yields. The performance of a bismaleimide in the NBS smoke chamber, shown in Figure 3.3-27 in the thermoplastics portion of this section, is significantly better than most other resins (maximum smoke density of 5–10 vs. 90–150 for epoxies). The char yield of PMR-15 is considerably better than that of the typical epoxy. Anaerobic char yields were reported [117] to be 66 for

PMR-15, 52–54 for PMR-II, and 34 for a DGEBA/Novolac cured with an aromatic diamine.

Commercial Availability. Polyimides are available in prepreg form from most of the vendors who supply aircraft and electrical circuit board markets. A representative (but not exhaustive) list of commercially available polyimide prepregs is found in Table 3.3-56. All three major categories (condensation, "addition," and bismaleimide) are available from the major suppliers.

3.3.6 Phenolic Resins

Introduction

Phenol-formaldehyde ("phenolic") resins are the oldest completely synthetic polymers, dating back to the 1907 patents of Leo H. Baekeland. Today, they are the most widely used of all the thermosetting resins; in 1981, they accounted for about 2.4 billion pounds or about 48% of all thermosetting resin consumption [1]. As a class, the phenolics are used in a wide variety of applications such as wood products, molding powders, and insulation. Laminate use is a relatively minor application for phenolics, accounting for about 7% of overall use. Most phenolic use is with fillers or fibrous reinforcements where the resin serves a binder or adhesive function and a small percentage of phenolic resin output goes into protective coatings.

Chemistry

Preparation. Although phenolic resins may be made from a wide variety of phenols and aldehydes, phenol

and formaldehyde account for most of the resin production. The minor starting materials for phenolics include cresols, alkyl phenols, paraformaldehyde, and hexamethylenetetramine (HMTA or HEXA).

Phenolics are usually divided into two classes: resoles and novolacs, also referred to as one-step and two-step resins, respectively. Resoles, or one-step resins, are typically prepared under alkaline conditions with formaldehyde/phenol (F/P) ratios of more than one. The reaction is stopped by cooling to give a reactive and soluble polymer which can be stored at low temperature and subsequently cross-linked by heat alone in what is essentially a continuation of the initial reactions, hence, the name "one-step." The resoles are the type of phenolic resin used in prepregs with papers or fabrics for both electrical and structural laminates. By contrast, novolacs are prepared under acidic conditions

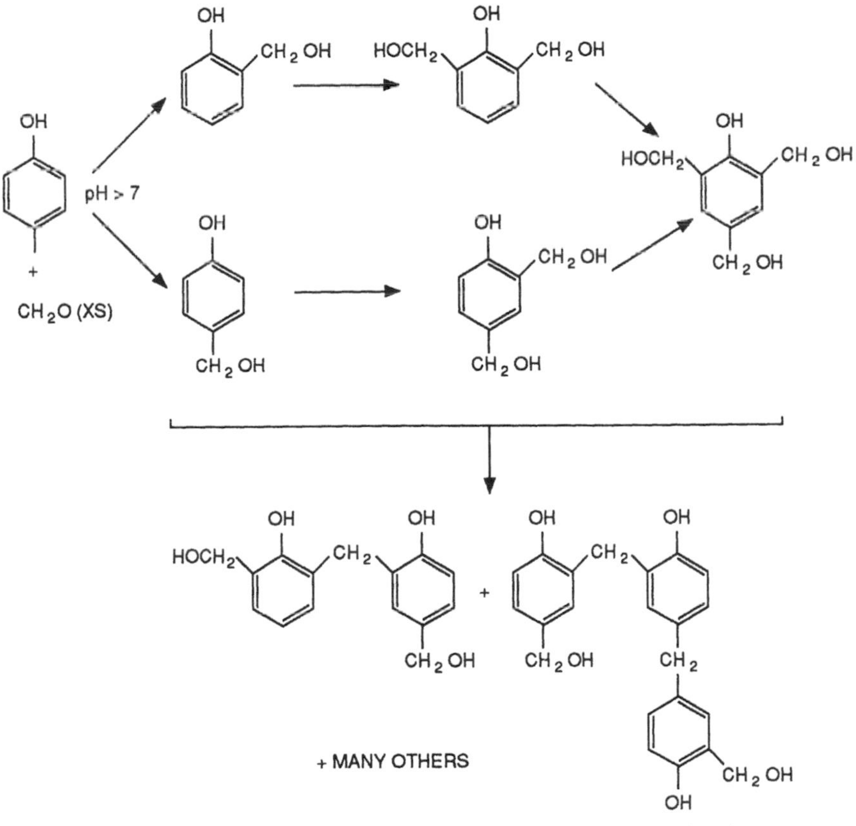

FIGURE 3.3-50. Initial structures formed in the preparation of resole phenolics.

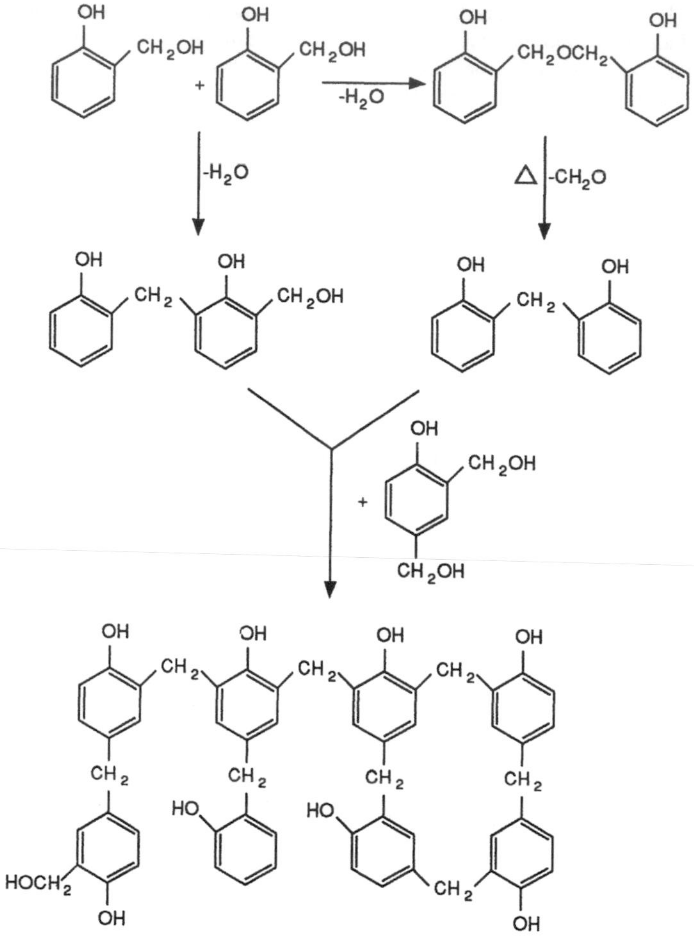

FIGURE 3.3-51. Reactions during cure of resole phenolics.

with an F/P ratio of less than one. The reaction is carried to completion to give an unreactive, thermoplastic oligomer which is dehydrated, pulverized, mixed with various fillers, and cured in a second step ("two-step") by the addition of an appropriate curing agent — usually hexamethylenetetramine. The novolacs are used in molding compounds, friction products, and other applications but are not used directly in fiber reinforced composites to any large extent.

The initial structures formed in resole preparation are shown in Figure 3.3-50; five of these structures are single-ring hydroxymethyl phenols (HMP), resulting from reaction of one, two, or three moles of formaldehyde with phenol at the *ortho* and *para* positions on the ring. The initially formed hydroxymethyl phenols may self-condense or react with free phenol to form multi-ring oligomers, as shown in Figure 3.3-50. The reaction is stopped at an appropriate point, and the solution of phenol alcohols and higher condensation products is used for prepregging or other applications. The composition of the resole is complex and is a function of several variables, including type of phenol used, type of catalyst, stoichiometry, time, temperature, and pH [151].

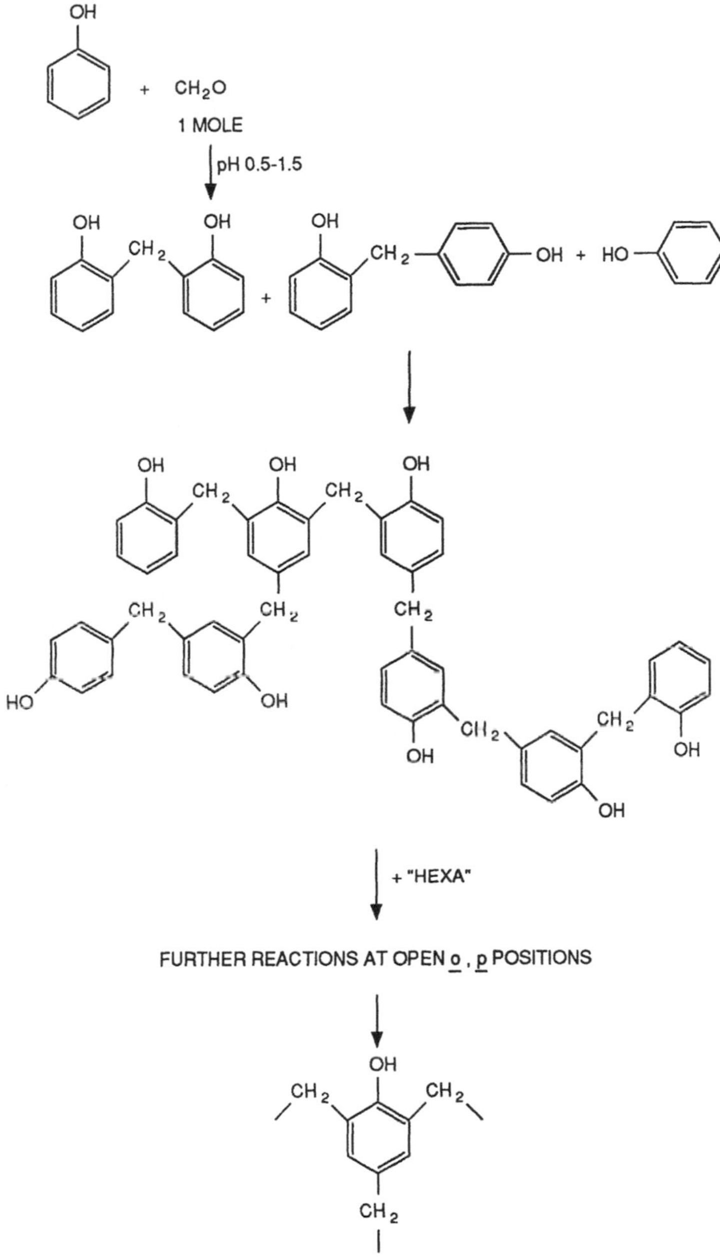

FIGURE 3.3-52.

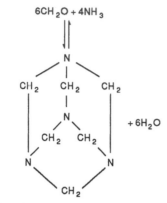

$$6CH_2O + 4NH_3$$

HEXAMETHYLENE TETRAMINE (HMTA)
("HEXA")

MW = 140
$(CH_2)_6 N_4$

SOLUBILITY: HIGHLY SOLUBLE IN WATER
SOLUBLE IN CHLOROFORM
SLIGHTLY SOLUBLE IN ALCOHOL

FIGURE 3.3-53. Preparation and structure of hexamethylene tetramine.

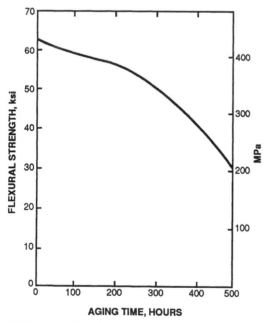

FIGURE 3.3-54. Flexural strength at 600°F vs. aging time at 500°F for *E*-glass fabric/SC-1008 phenolic laminates.

Curing of Resole Phenolics. Resoles are cured by heat or by addition of acids (or both) [152]. Heat cure in the 130–200°C range is by far the most important method used in composite applications. The cure reactions are condensations, which liberate one mole of water (or formaldehyde) for each new bond formed. The chemistry of the curing process is complex because there are many structures present and many competing reactions—each influenced by reaction conditions and other variables. Only a few important aspects of cure will be covered here. More detailed discussions of the curing process are given in references [151,152].

The two most important curing reactions are shown in Figure 3.3-51: the first one is the reaction of a methylol group with a range hydrogen (*ortho* or *para*) on another molecule to give a methylene bridge between two rings, and the second one is the reaction of two methylol groups to form a dibenzyl ether bridge between two rings. In thermal cures below 170°C, these two reactions occur simultaneously [151]. At higher temperatures the ether link is converted to a methylene link by loss of formaldehyde, as shown in Figure 3.3-51. The methylene link is the more thermodynamically stable product and is regarded as the most important linkage developed during cure [151,152].

Cure of resoles can also occur at low temperatures if acidic catalysts (e.g., p-toluenesulfonic acid) are used, but this route is of limited commercial importance [152].

Preparation and Cure of Novolacs. The preparation of novolacs is outlined in Figure 3.3-52. Here the pH is low and phenol is in excess, so the products contain no methylol groups and are unreactive until mixed with a hardener that provides methylene groups (e.g., hexamethylenetetramine). The predominant bridging gap is methylene, both in the thermoplastic stage and in the cured resin [153]. The molecular weight, although higher than for resoles, is still relatively low; number-average molecular weights of about 900 and weight-average molecular weights of 5,000 have been reported for a "high molecular weight" novolac [154].

Since they are unreactive alone, novolacs are cured by the addition of a second component which typically serves as a methylene donor to provide the bridging groups to cross-link the polymer. These groups react at the available *ortho* or *para* positions on the rings, shown in Figure 3.3-52, to form the network. Although a number of curing agents may be used, hexamethylenetetramine (HEXA) is the most common; its prepa-

Table 3.3-57. *Characteristics of an unmodified state-of-the-art heat-resistant phenolic.*

(RESINOX SC-1008)	
Uncured	
% solids	60–64
Viscosity, cps	180–300
pH	7.9–8.5
Density, g/cm³	1.07–1.10
Cured Resin Properties	
Tensile Strength, psi	8,900
Tensile Modulus, 10³ psi	570
Tensile Elongation, %	1.7
Shear Modulus, 10³ psi	240
Poisson's Ratio	0.375
Density, g/cm³	1.26

ration, from formaldehyde and ammonia, is shown in Figure 3.3-53. Ammonia is a by-product in cures with HEXA. Other curing agents include paraformaldehyde, trioxane, and resole phenolics.

Properties

The properties prior to and after cure of an unmodified phenol-formaldehyde resole are shown in Table 3.3-57. Tensile strength and modulus values are very similar to those of structural epoxies based on MY-720.

The high temperature aging performance of this resin in a glass fabric laminate is shown in Figure 3.3-54, where flexural strength at 500°F is plotted vs. aging time at 500°F; after 500 hours of aging, the flexural strength is almost 50% of its initial 500°F value and is 38% of its initial room temperature value.

Phenolics are noted for good thermal stability, high char yields, and low smoke generation [152,154]. Up to about 300°C, the phenolic polymer remains intact, and weight loss is due to water and monomers diffusing out. About 300°C degradation, which is claimed to be thermo-oxidative whether carried out in air or an inert atmosphere, is attributed to the high oxygen content of the resins [155]. The methylene linkage is, as discussed previously, the major cross-link in the cured resin and the site most susceptible to oxidative attack, which leads to similar degradation behavior among different, but well-cured, resins [152].

Smoke performance for unmodified phenolics is outstanding in typical NBS Smoke Chamber tests; typical maximum smoke density values of about 5 were noted for glass/phenolic laminates [154]. Anaerobic char yields of 48–61% have been reported [154,156]. In both these references, the phenolic was an unmodified phenol-formaldehyde resin. Many commercial phenolic prepregs give smoke density values in the 10–50 range because they have been modified to achieve flow control in co-cured sandwich panel constructions; the typical

Table 3.3-58. *Representative commercial phenolic prepregs.*

Source	Designation	Description
Ciba-Geigy	R 917	Modified
Ferro	Pyropreg AC	Modified
Ferro	CPH-2209	High temp., high strength; MIL-R-9299
Ferro	CPH-2251	Low flammability and smoke
Ferro	CPH-2280	Low free phenol; str. retent. to 500°F; self-adhesive
Ferro	CPH-2287	New
Fiberite	6055	Structural
Fiberite	6065	Structural for commercial/industrial
Fiberite	6070	Press grade interior
Fiberite	6800	High peel str., vac-bag grade, interiors
American Cyanamid	CYCOM 6102-2	Low pressure curing, 350°F cure
American Cyanamid	CYCOM 6102	High tack, low pressure, 350°F cure
American Cyanamid	CYCOM 6113	Aircraft interior grade, 325°F cure
American Cyanamid	CYCOM 6200	Quick cure, crushed core, 310°F cure
American Cyanamid	CYCOM 6142	Low temp. cure (250°F), interiors
American Cyanamid	CYCOM 6162	High peel, low temp. cure (250°F), aircraft int.

modifiers are thermoplastic polymers with poorer smoke and flammability performance than the base phenolic resin.

Applications and Commercial Availability

For composite applications, prepregs are a major form in which phenolics are sold and used. A wide variety of phenolic prepregs is available: a representative selection is shown in Table 3.3-58. Although not stated directly in each case, low smoke and good flammability resistance are major reasons for most of the uses. A majority of the listed prepregs are intended for aircraft interior applications of various types, and several are formulated for co-cured sandwich panel construction. Co-cured panels are prepared by curing and adhering the facing to the core in one operation. Cure temperatures can be as low as 250°F and are usually in the 250–350°F range. For parts where high temperature performance is needed, a postcure at or above the use temperature is usually recommended.

A phenolic sheet molding compound is also available commercially [157]. This material is useful for molding complex shapes where fiber flow is needed; it can be molded in tools designed for polyester SMC and on similar cycles. Comparative smoke density values are 8 for the phenolic SMC vs. 107 for a polyester control.

3.3.7 References

1. SPI Committee on Resin Statistics as Compiled by Ernst & Ernst.
2. BOENIG, H. V. *Unsaturated Polyesters: Structure and Properties*. Elsevier, Amsterdam (1964).
3. Stanford Research Institute, *Chemical Economics Handbook*.
4. KROEKEL, C. H. Can. Patent 820,399.
5. VANCSO-SZMERCSANYI, I. et al. *J. Polymer Sci.*, 53:241 (1961).
6. PARK, R. E., R. M. Johnston, A. D. Jesensky and R. D. Cather. SPI/RP/C, 17th Ann. Tech. Conf. (1962).
7. Amoco Chemicals Corp. Bulletin IP-69.
8. SMITH, A. L. and J. R. Lowry. SPI, Reinforced Plastics Div. (1960).
9. OLEESKY, S. S. and J. G. Mohr. *Handbook of Reinforced Plastics*. Van Nostrand, New York, NY, p. 37 (1964).
10. SHEPPARD, C. S. and V. R. Kamath. SPI/RP/C, 33rd Annual Tech. Conf., Paper 5-D (1978).
11. PARKYN, B., F. Lamb and B. V. Clifton. *Polyesters, Vol. 2*. ILIFFE Books, Ltd., London, p. 86 (1967).
12. PASTORINO, R. L. In *Unsaturated Polyester Technology*, P. F. Bruins, ed., Gordon & Breach Science Publishers, New York, NY, p. 63 (1976).
13. LONGENECKER, D. M. In *Unsaturated Polyester Technology*. P. F. Bruins, ed., Gordon & Breach Science Publishers, New York, NY, p. 279 (1976).
14. YOUNG, R. E. In *Unsaturated Polyester Technology*. P. F. Bruins, ed., Gordon and Breach Science Publishers, New York, NY, p. 315 (1976).
15. REGESTER, R. F. *Corrosion*, 25(4):157 (1969).
16. FRIED, N. In *Mechanics of Composite Materials*. F. W. Wendt, H. Liebowitz and N. Perrone, eds., Pergamon, Oxford, p. 814 (1970).
17. ALLEN, R. C. SPI/RP/C, 33rd Ann. Tech. Conf., Paper 6-D (1978).
18. ISHIDA, H. and J. L. Koenig. *Polymer Eng. Sci.*, 18(2):128 (1978).
19. REGESTER, R. F. SPI/RP/C, 22nd Ann. Tech. Conf., Paper 16-D (1967).
20. ASHBEE, K. H. G. and R. C. Wyatt. *Proc. Roy. Soc.*, A312:553 (1969).
21. CELLITTI, C., S. Varigu, E. Martuscelli and D. Fabbro. *J. Materials Sci.*, 22:3477 (1987).
22. Dow Chemical Co. Bulletin, "Derakane Vinyl Ester Resins, Chemical Resistance Guide," Form No. 190-273-77 (1977).
23. GOOLD, R. D., T. S. McQuarrie and R. Obrycki. SPI/RP/C, 35th Ann. Tech. Conf., Paper 14-E (1980).
24. BEITIGAM, W. V. SPI/RP/C, 33rd Ann. Tech. Conf., Paper 6-E (1978).
25. MILLER, D. P., R. V. Petrella and A. Manca. SPI/RP/C, 31st Ann. Tech. Conf., Paper 20-C (1976).
26. TRAMPENAU, R. H. and T. R. Evans. SPI/RP/C, 28th Ann. Tech. Conf., Paper 4-C (1973).
27. BONSIGNORE, P. V. and J. H. Manhort. SPI/RP/C, 29th Ann. Tech. Conf., Paper 23-C (1974).
28. CHANDIK, B. V. *Modern Plastics*, p. 72 (October 1977).
29. GOINS, O. K. and B. V. Chandik. SPI/RP/C, 31st Ann. Tech. Conf., Paper 3-A (1976).
30. BARRENTINE, E. M., R. D. Goold, F. Roselli and U. Toggweiler. SPI/RP/C, 24th Ann. Tech. Conf., Paper 16-B (1969).
31. STAVINOHA, R. F., L. Craigie, T. F. Anderson and L. Hartless. SPI/RP/C, 31st Ann. Tech. Conf., Paper 3-B (1976).
32. BATZER, H. and S. A. Zahir. *J. Appl. Polymer Sci.*, 19:585 (1975).
33. HABERMEIER, J. *Angew. Makromol. Chem.*, 35:9 (1974).
34. Ciga-Geigy Tech. Data Bulletin. "Developmental Resin XB2793," Ardsley, NY (1978).
35. PEREZ, R. J. In *Epoxy Resin Technology*. P. F. Bruins, ed., Interscience, New York, NY, p. 45 (1968).
36. TANAKA, Y. and T. F. Mika. In *Epoxy Resins*. C. A. May and Y. Tanaka, eds., Marcel Dekker, New York, NY, Chap. 3 (1973).

37. HARRIS, J. J. and S. C. Temin. *J. Appl. Polymer Sci.*, 10:523 (1966).

38. TANAKA, Y. and H. Kakiuchi. *J. Appl. Polymer Sci.*, 7:1063 (1963).

39. LEE, H. and K. Neville. *Handbook of Epoxy Resins.* McGraw-Hill, New York, NY (1967).

40. SCHECHTER, L., J. Wynstra and R. P. Kurkjy. *Ind. Eng. Chem.*, 48(1):94 (1956).

41. BELL, J. P. *J. Polymer Sci.*, A-2(6):417 (1970).

42. ANDERSON, H. C. *SPE Journal*, 16:1241 (1960).

43. HORIE, K. *J. Polymer Sci.*, A-1(3):1357 (1970).

44. POGANY, G. A. *Polymer*, 11:66 (1970).

45. ARRIDGE, R. G. C. and J. H. Speake. *Polymer*, 13:450 (1972).

46. PATTERSON-JONES, J. C. *J. Appl. Polymer Sci.*, 19:1539 (1975).

47. CONLEY, R. T. In *Thermal Stability of Polymers, Vol. I.* R. T. Conley, ed., Chap 11 (1970).

48. SCHECHTER, L. and J. Wynstra. *Ind. Eng. Chem.*, 48(1):86 (1956).

49. BRESLAU, A. J. *Epoxy Resins.* C. A. May and Y. Tanaka, eds., Marcel Dekker, New York, NY, Chap 8 (1973).

50. Celanese Resins, Tech. Bulletin 0366.

51. CHIAO, T. T., E. S. Jessop and H. A. Newey. *SAMPE Quarterly*, 6(1):28 (1974).

52. RINDE, J. A., E. R. Mones and H. A. Nervey. Univ. of Calif. Pub. No. UCRL-52577 (1978).

53. Dow Chemical Co. Bulletin *D.E.R. 383 Epoxy Resin.*

54. DAY, R. D. *J. Reinforced Plastics and Composites*, 7:475 (1988).

55. VARMA, I. K. and P. V. Satya Bhama. *J. Composite Materials*, 20:410 (1986).

56. DUSI, M. R., W. I. Lee, P. R. Ciriscioli and G. S. Springer. *J. Composite Materials*, 21:243 (1987).

57. LONG, E. R. NASA Tech. Paper 1474 (N79-29241) (August 1979).

58. AUGL, J. M. NSWC TR-79-39 (AD-A076008).

59. SPRINGER, G. S. and C. H. Shen. AFML-TR-76-102.

60. *Thermoplastic Composite Materials.* L. A. Carlsson and R. B. Pipes, eds., Elsevier, Amsterdam (1989).

61. HOGGATT, J. T. Advanced Fiber Reinforced Thermoplastics Program, Contract F33615-76-C-3048 (December 1976).

62. HOGGATT, J. T. Advanced Fiber Reinforced Thermoplastics Program, Contract F33615-76-C-3048 (April 1977).

63. JOHNSTON, N. J. and P. M. Hergenrother. *Proc. 32nd Intl. SAMPE Symposium.* p. 1412 (April 6–9, 1987).

64. *Chemical and Engineering News* (February 22, 1982).

65. TING, R. Y. *J. Materials Sci.*, 16:3059 (1981).

66. WANG, A. S. D., et al. In *Mechanical Behavior of Materials, Vol. 2.* Proceedings, 1974 Symposium, Soc. of Material Sciences, Kyoto, Japan (1974).

67. ESCH, J. G., et al. *Soc. Plastics Ind.*, RP/C Conf., Paper 15-G (1983).

68. *Modern Plastics.* p. 60 (May 1974).

69. HUSMAN, G. E. and J. T. Hartness. *SAMPE National Symposium*, 24:21 (1982).

70. HARTNESS, J. T. *SAMPE National Technical Conference*, 14:26 (1982).

71. HARTNESS, J. T. *SAMPE National Symposium*, 25:376 (1980).

72. KNIGHT, G. J. In *Developments in Reinforced Plastics—1.* G. Pritchard, ed., Applied Science Pub., London, p. 145 (1980).

73. PATTERSON-JONES, J. C. *J. Appl. Polymer Sci.*, 19:1539 (1975).

74. NOVAK, R. C. NASA CR-134881 (July 1975).

75. ICI Petrochemicals and Plastics Div., Provisional Data Sheet PK PD17 (1981).

76. NELSON, G. L. *J. Fire and Flammability*, 5:125 (1974).

77. MEGAL, C. A., et al. In *International Symposium on Flammability and Fire Retardance.* V. M. Bhatnagar, ed., Technomic Publishing Co., Inc., Lancaster, PA (1977).

78. KOURTIDES, D. A. and J. A. Parker. *SAMPE National Symposium*, 23:893 (1978).

79. WITTENWYLER, C. V. Fire Retardant Chemicals Assoc., Conf./Workshop (March 1978).

80. VAN KREVELEN, D. W. *Polymer*, 16:615 (1975).

81. MAYOR, R. A. and A. Ossin. *SAMPE National Symposium*, 24:1567 (1979).

82. HOGGATT, J. T. and A. D. Von Volkli. AD-A011407 (March 1975).

83. MAY, L. C. and R. C. Goad. *Manufacturing Methods for Fabrication and Assembly of Advanced Composite Primary Aircraft Structure*, AFML-TR-75-111 (July 1975).

84. SHEPPARD, C. H. and E. E. House. *SAMPE National Tech. Conf.*, 14:70 (1982).

85. HOGGATT, J. T., AD-763470 (June 1973).

86. HOGGATT, J. T. Advanced Fiber Reinforced Thermoplastics Program, Contract F33615-76-C-3048 (July 1977).

87. HOGGATT, J. T. Advanced Fiber Reinforced Thermoplastics Program, Contract F33615-76-C-3048 (January 1979).

88. JAQUISH, T., et al. *Graphite Reinforced Thermoplastic Composites*, AD-A102217 (August 1980).

89. WIZLER, S. *Advanced Composites*, 3(5):49 (1988).

90. NOVAK, R. C. NASA CR-135196 (April 1977).

91. HARTNESS, J. T. and J. A. Dyksterhouse. *SAMPE National Tech. Conf.*, 13:514 (1981).

92. ICI Petrochemicals and Plastics Div., Provisional Data Sheet APC PD3.

93. LISKAY, G. G. *SAMPE National Technical Conference*, 13:592 (1981).

94. NOLET, S. C. and J. P. Fanucci. *Proceedings from a Workshop on Design, Manufacture and Quality Assurance of Low Cost, Light Weight Advanced Composite Molds, Tools*

and Structures, Florida Atlantic University, Boca Raton, FL (October 17–19, 1988).

95. Advanced Materials Newsletter, Vol. 4, No. 20 (November 22, 1982).

96. HOUSE, E. E. SAMPE Nat. Sym., 24:201 (1979).

97. COOKE, L. SAMPE, Orange County Seminar, Anaheim, California (February 10, 1983).

98. HILL, S. G. and M. Stander. "Environmental Exposure of Thermoplastic Composites," unpublished report (May 1978).

99. HERGENROTHER, P. M. J. Polymer Sci., 20:3131 (1982).

100. HILL, S. G., E. E. House and J. T. Hoggatt. Advanced Thermoplastic Composite Development, Naval Air Systems Command, Contract N00019-77-C-0561 (May 1979).

101. YOUNG, et al. AFFDL Report TR-78-109 (August 1978).

102. BYERS, B. A. NASA CR-159293 (August 1980).

103. SHEPPARD, C. H. and E. E. House. Graphite Reinforced Thermoplastic Composites, NASC Contract N00019-80-C-0365 (December 1981).

104. HILL, H. W., Jr. In Durability of Macromolecular Materials. R. K. Eby, ed., ACS Symposium Ser. 95, p. 185 (1978).

105. LONG, W. C. Reinforced Thermoplastics, AMMRC Report CTR-74-32 (April 1974).

106. WHITE, S. A. et al. J. Appl. Polymer Sci., 27:2675 (1982).

107. BURKS, H. D. and T. L. St. Clair. NASA Technical Memorandum 844494 (May 1982).

108. ST. CLAIR, A. K. and T. L. St. Clair. SAMPE National Symposium, 26:165 (1981).

109. ICI Petrochemicals and Plastics Div., Provisional Data Sheet APC PD2.

110. WITZLER, S. Advanced Composites, 3(2):55 (1988).

111. SRI Chemical Economics Handbook (December 1980).

112. SROOG, C. E. J. Poly. Sci., C16:1191 (1967).

113. SERAFINI, T. T. and P. Delvigs. Applied Polymer Symposia, 22 (1973).

114. KOURTIDES, J. J. Thermoplastic Composite Materials, 1:12 (1988).

115. ALVAREZ, R. T. and F. P. Darmory. SPE/ANTEC, p. 687 (May 1974).

116. Rhône-Poulenc Bulletin K/IE/783, Kerimid (July 1983).

117. CAVANO, P. J. NASA CR-159666 (October 1979).

118. LAUVER, R. W. J. Poly. Sci., 17:2529 (1979).

119. KRAY, R. J. NASA CR 3276 (April 1981).

120. ST. CLAIR, T. L. and R. A. Jewell. SAMPE Nat. Sym., 23:520 (1978).

121. DELOS, S. E., et al. J. Appl. Polymer Sci., 27:4295 (1982).

122. YOUNG, P. R. and G. F. Sykes. In Resins for Aerospace. C. A. May, ed., American Chemical Society, Washington, DC (1980).

123. MACE, W. C., et al., SAMPE Nat. Sym., 24:217 (1979).

124. BILOW, N., et al. SAMPE Nat. Sym., 23:791 (1978).

125. BILOW, N., et al. SAMPE Journal, 18(1):8 (1982).

126. APONYI, T. J., et al. SAMPE Nat. Sym., 23:763 (1978).

127. DYNES, P. J., et al. J. Appl. Polymer Sci., 25:1059 (1980).

128. GIBBS, H. H. SAMPE Nat. Sym., 29:1073 (1984).

129. Upjohn Co., Report No. 5, "Polyimide 2080."

130. ST. CLAIR, T. L. and A. K. St. Clair. SAMPE Quarterly, p. 20 (October 1981).

131. GIBBS, H. H. SAMPE Journal, p. 37 (Sept./Oct. 1984).

132. HARRUFF, P. W. and P. R. Scherer. SAMPE Nat. Tech. Conf., 11:1 (1979).

133. DELVIGS, P., W. B. Alston and R. D. Vannucci. SAMPE Nat. Sym., 24:1053 (1979).

134. JONES, J. S. SAMPE Nat. Tech. Conf., 14:402 (1982).

135. WILKINS, D. J. AD-A 112474 (September 1981).

136. ODOM, E. M. and D. F. Adams. NADC-83053-60 (June 1983).

137. BASCOM, W. D., et al. NRL Memorandum Report 4005 (May 1979).

138. MA, C. C. M., H. C. Hsia, W. L. Liu and J. T. Hu. J. Thermoplastic Composite Materials, 1:39 (1988).

139. DAVIS, J. G., JR. NASA Conf. Pub. 2142 (August 1980).

140. WRIGHT, W. W. In Developments in Polymer Degradation—3. N. Grassie, ed., Applied Science, London (1981).

141. STREET, S. W. SAMPE Nat. Sym., 25:366 (1980).

142. PATER, R. H. SAMPE Journal, p. 17 (Nov./Dec. 1981).

143. KERR, J. R. and J. F. Haskins. AIAA Journal, 22(1):96 (1984).

144. DINE-HART, R. A., et al. British Polymer J., 3:222 (1971).

145. DEIASI, R. Grumman Research Dept. Memorandum RM-589 (August 1974).

146. MAURI, R. E., F. W. Crossman and W. J. Warre. SAMPE Nat. Sym., 23:1202 (1978).

147. MOOREFIELD, S. A., J. B. Styron and L. C. Hoots. ARML-TR-77-182.

148. U.S. Patent 4,288,359.

149. French Patent Fr 2,476,102.

150. Japanese Patent 81,104,924.

151. MARTIN, R. W. The Chemistry of Phenolic Resins. John Wiley & Sons, New York, NY (1956).

152. KNOP, A. and W. Scheib. Chemistry and Applications of Phenolic Resins. Springer-Verlag, New York, NY (1979).

153. BRODE, G. L. "Phenolic Resins," in Encyclopedia of Chemical Technology, 3rd edition (1982).

154. KOURTIDES, D. A. and J. A. Parker. SAMPE Nat. Symp., 23:893 (1978).

155. CONLEY, R. T. Thermal Stability of Polymers. Marcel Dekker, New York, NY, Chapter 11 (1970).

156. DELANO, C. B. and A. H. McLeod. "Synthesis of Improved Phenolic Resins," NASA CR-159724 (1979).

157. Reichhold Chemicals, Inc., Tacoma, Wash., Bulletin No. TFP-12, 2-83.

SECTION 3.4

Molding Processes— An Overview

3.4 MOLDING PROCESSES — AN OVERVIEW

M. G. BADER

3.4.1 Summary of Basic Types

For the purposes of this brief survey, molding processes for reinforced plastics will be classified under open mold, matched-die, SMC, BMC, compression, transfer, injection, reaction injection, press, autoclave, and pultrusion processes. The aim is to set out a brief comparison of the alternatives with the scope of each process indicated so that this section may be used as a first stage in the process of selecting a suitable fabrication route for a given application. Beginning with section 3.5, these processes will be taken up in greater detail. The design and manufacture of molds will be discussed only briefly. For more detail, see the recent book by Morena [1].

Open Mold Processes

As the title suggests, these processes are carried out in open molds, which may be of male or female form—usually the latter. In the basic "contact molding" process (Figure 3.4-1), a mold is prepared from a master pattern, and the reinforced plastic is applied to the mold to reproduce the shape of the original. Thus, for a typical part, say a small boat hull, the master might be made from wood and accurately finished, varnished, and polished to produce a finish as good as that required on the molding. The mold would then be made from GRP by laminating chopped strand mat and/or woven rovings with a suitable resin onto the master, making the necessary constructions for multiple part molds according to the design requirements. The molding sequence is illustrated in Figure 3.4-1.

First, the mold surface is polished and treated with a mold release agent to prevent the molding from sticking to the mold. Then a coating of resin—often pigmented, called the gel coat—is applied to the mold surface. This resin will become the prime surface of the molding; the quality of the surface is determined by the care with which this gel coat is applied. In glass/polyester systems, it is usual to use a special thixotropic resin for the gel coat. This gel coat is applied to give a coating of 0.5–1.0 mm (0.02–0.04 in). It may be applied by brush, roller, or spray. When the gel coat has set, a back-up lamina of a fine glass-surfacing tissue is often applied, followed by layers of chopped strand mat (CSM) and/or woven roving and resin according to the design specifications. In the basic process, these are all manual operations. For each lamina, a coating of resin is first applied by brush or spray, then by the dry mat or cloth. The resin is worked into the cloth, usually with rollers, until the layer is uniformly wetted out and impregnated with resin. Each layer is usually allowed to gel but not to cure completely between each application. From this description it can be seen that the process is labor-intensive and that overall quality depends entirely on manual skills.

When the final layer has been applied, the whole laminate is allowed to cure, usually at room temperature; it is then trimmed where necessary, and broken from the mold, which may then be prepared for the next molding operation.

This process is most commonly used with room temperature cure polyester resins and glass reinforcement but can be used also with vinyl ester and suitable epoxy systems.

The production rate is determined by the manpower that can be deployed on an individual molding, the lay-up time, and the cure time. Fast-curing resins have short pot lives so that a constant supply of freshly catalyzed resin is required. Production rates may be increased by pre-tailoring the reinforcement, but in many cases it is applied directly from the roll and cut to length as required.

The process survives because it is very adaptable and low quality labor can readily acquire the necessary skills. Because it is slow, it is mainly applicable to the production of relatively large moldings, where not

89

BASIC COLD-CURE CONTACT MOLDING PROCESS

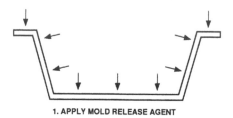

1. APPLY MOLD RELEASE AGENT

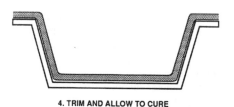

2. APPLY GEL COAT

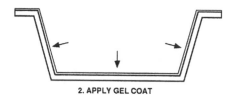

3. APPLY REINFORCEMENT (MAT AND/OR CLOTH) + RESIN - WET-OUT AND COMPACT

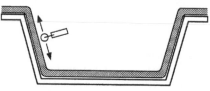

4. TRIM AND ALLOW TO CURE

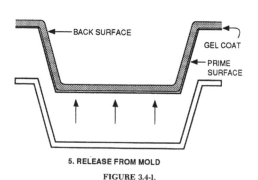

5. RELEASE FROM MOLD

FIGURE 3.4-1.

SPRAY-UP CONTACT MOLDING PROCESS

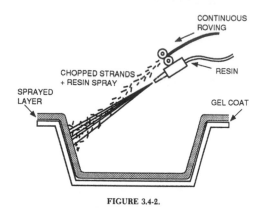

FIGURE 3.4-2.

more than one molding per mold per working day is practicable. There is virtually no limitation to the size of molding. For larger moldings, one can dispense with the master form and build the molds directly in steel, or even concrete, using templates or generated curves. Among the largest moldings produced to date are 84 m (275 ft) hulls for minehunters for the British Royal Navy. These were produced in steel female molds from 1 m (39 in) wide sheets of woven glass rovings and polyester resin. Parts of the hull are over 50 mm (2 in) thick and comprise over 100 layers of woven roving. The molding operation took several months, and a special resin formulation, having a long gel time, was used. This gel time allowed a period of several days between the application of successive laminae. At the other end of the scale, small moldings produced in great numbers may also utilize metal molds, which have a longer life than the traditional GRP.

The process can be mechanized in a number of ways. First, the reinforcement can be preimpregnated with resin before being applied to the mold. This process reduces the laminating period and is applicable to large moldings. The "spray-up" process is a method in which resin is sprayed onto the mold at the same time as the chopped rovings (Figure 3.4-2). This combination is sprayed with a special applicator that incorporates a resin spray and chopping head which, of course, is suitable for random fiber construction only when the glass content of the molding is generally rather low. However, this spray application is balanced by a greater degree of control and faster operation. The sprayed-on fiber and resin still need to be finally consolidated by hand-rolling, but the amount of labor is much reduced. The resin is mixed and metered automatically; thus, a more consistent quality can be maintained.

In some cases final cure can be accelerated by heating. Small molds can be passed through air circulating ovens or radiant heat tunnels. Larger moldings can be heated by batteries of infrared heaters. It is even possible to incorporate heating foils in the laminate to assist cure.

A further possibility is to exert pressure on the curing laminate by means of a vacuum applied between the molding surface and a suitable rubbery membrane material. This vacuum can be applied only to moldings of a suitable form but provides better consolidation of the laminate and improves the quality of the "second" surface.

Matched-Die Processes

A number of distinct processes fall within this classification. The common feature is a two-part mold or die which forms a cavity corresponding to the shape of the article to be molded (Figure 3.4-3). The molds are most commonly made from metal (steel, cast iron, zinc alloy, or aluminum alloy), and high temperature cure is usual. The molding operation is also usually carried out while pressure is applied to molds, which assures that the mold halves are properly in register so that accurate moldings are produced. In contrast to the contact molding processes, all surfaces of the molding can have a smooth finish.

Resin Injection Molding

This is a relatively new process for manufacturing large moldings in GRP. There are several variants of the process. Molds are generally of a form similar to those used for contact molding but are rather more rigidly constructed in the form of a matched pair or, perhaps, a multipiece mold. The reinforcement, consisting of carefully tailored woven roving, cloth, and/or mat, is laid in position (dry) in the female cavity. When all the reinforcements, including sandwich core materials and stiffening rib formers, are in place, the male half of the mold is placed in position—usually by bolting the two

BASIC MATCHED-DIE MOLDING PROCESS

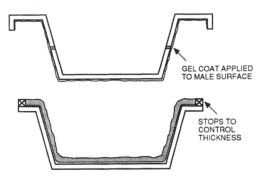

1. **FEMALE MOLD WITH GEL COAT AND REINFORCED LAYERS PREPOSITIONED**

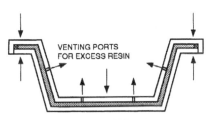

2. **MOLD CLOSED TO STOP — RESIN ALLOWED TO CURE**

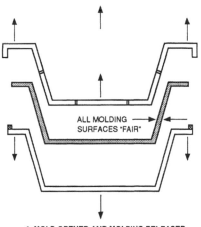

3. **MOLD OPENED AND MOLDING RELEASED**

FIGURE 3.4-3.

RESIN INJECTION PROCESS

**1. BOTH MOLDS GEL COATED IF REQUIRED —
REINFORCEMENT PLACED DRY IN FEMALE MOLD**

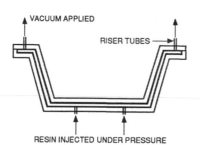

**2. MOLD HALVES CLOSED — RESIN INJECTED AT BOTTOM —
VACUUM APPLIED AT TOP — RISER TUBES PINCHED OFF
AS RESIN RISES AND IS ALLOWED TO CURE —
THEN MOLD OPENED AND PART RELEASED**

FIGURE 3.4-4.

COMPRESSION MOLDING

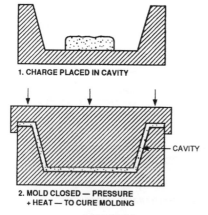

1. CHARGE PLACED IN CAVITY

**2. MOLD CLOSED — PRESSURE
+ HEAT — TO CURE MOLDING**

FIGURE 3.4-5.

halves together along a mating flange. The mold may be further reinforced by means of an external metal framework. Liquid resin is now admitted into the mold, usually through several ports at the bottom of the mold. The infiltration of the resin is assisted by drawing a vacuum on the cavity from the top of the mold or by injecting the resin under positive pressure or both (see Figure 3.4-4). The key to success is that the rate of flow of the resin into the mold should match the rate at which the reinforcement is wetted out. The rate must also be carefully controlled so that the resin rises uniformly from bottom to top of the molding, to prevent formation of dry areas [2,3].

This can be a long process, taking from several hours in the case of the largest moldings [10–20 m (30–60 ft) boat hulls] to a few minutes for small moldings. Clearly, a resin of low viscosity and relatively long gel time is required. The advantage of the process is that the reinforcement is precisely located, the resin content can be accurately controlled, and an excellent finish may be obtained on all surfaces. If a gel coat is required, it may be applied to the molds before the reinforcement is placed. For the larger moldings where GRP molds are used, a cold-curing resin must be used, e.g., polyester or vinyl ester. The principle may also be applied to smaller moldings required in large numbers. In this case metal molds are used—the reinforcement is placed as before, but the molds are then closed in a press and the resin injected, usually under a low positive pressure and with vacuum assistance to reduce formation of voids. The metal molds are usually heated to achieve a rapid cure and a high production rate. The process is applicable to medium-sized moldings; polyester, vinyl ester, or epoxy resins can be employed.

For large moldings and for high production rate smaller moldings, it is convenient to add the cure promoters to the resin just before injection. In one process, equal quantities of resin blended with catalyst and accelerator are pumped to the molding—the two streams are mixed as they enter the mold. An alternative technique is to add a metered quantity of liquid catalyst to the pre-accelerated resin as the mold is filled. The former technique is generally more convenient because the catalyst is used only to the extent of a few percent of the resin weight and good catalyst distribution is easier to achieve. These techniques are similar to the RIM process discussed later. This process is sometimes referred to as resin transfer molding (RTM).

SMC, DMC, and BMC Molding

These processes differ radically from those discussed above in that the feedstock is a pre-blended mixture of reinforcement, thermosetting resin, and, usually, a filler. A pair of heated matched metal dies is used in a suitable press. The mold is opened and a pre-determined charge of material is placed in position. (The distribution of this charge is critical to the success of this process.) The mold is then closed under pressure, which forces the charge into the mold cavity. Excess charge is vented at the flow extremities of the molding. The charge cures rapidly in the hot mold, which is then opened and the finished molding ejected while still hot. Molding cycles of 2 minutes are quite common, and development is under way of systems that cure in ~30 seconds. The only fundamental difference between the SMC, DMC, and BMC materials, as far as processing is concerned, is in the form of the feedstock. The SMC is in a sheet of ~6 mm (0.25 in) thick, while DMC and BMC are usually in the form of a "rope" 20–50 mm (1–2 in) dia. BMC can also be handled in a dispenser system, which extrudes a metered charge directly into the mold cavity. Otherwise, the charges are normally prepared manually by weighing out the correct quantity of material and placing it in the mold according to the production schedule. The successful operation of this process depends on a uniform flow of material in the mold so that the reinforcing fibers are evenly distributed. Improper attention to charge placement can result in fiber-depleted regions with consequent loss of properties.

Compression Molding

In principle, compression molding is similar to the SMC process except that it is usually applied to resins such as phenolics, which require much higher consolidation pressures. In the SMC and allied processes, the molding pressure serves to close the mold to fixed stops. The material is distributed as the mold closes [4], and any excess is allowed to spill out of the mold. However, in compression molding the mold is designed to maintain continuous pressure on the charge, and the dimensions of the molding are determined by the accurate control of the charge weight. The feedstock for typical fiber reinforced thermosets is a dry, coarse powder. This is usually pre-compressed into pellets of a suitable charge weight. (One or more pellets may be used to form an individual charge.)

The pellets are dried and preheated, typically to 70–90°C. The appropriate charge is placed in the opened mold, which is then closed; pressure is applied, and the resin is allowed to cure—typically for 1–10 minutes; the mold is reopened and the finished molding ejected (Figure 3.4-5). The process is most commonly used for the mass production of relatively small intricate parts. Multicavity tools are very common, and metal inserts are often molded into the components. The quality of the molding is determined by proper mold design, control of the molding temperature cycle, and application of pressure in the correct sequence. Ideally the pressure is applied slowly as the charge softens but before it starts to gel. It is also essential that the mold be adequately vented to allow water vapor and other volatiles to escape during the curing process.

Transfer Molding

Transfer molding is a variant of compression molding and is applicable to similar materials although the flow requirement of the resin is more critical. The mold consists of two cavities—a true mold cavity and a transfer cavity, which is simply a cylinder in which a ram operates. The charge is placed in the transfer cavity, and, when it has softened, pressure is applied to the ram, causing the charge to flow through the transfer port into the mold cavity (Figure 3.4-6). The pressure on the ram is also generally employed as the mold closing force. An excess of charge is placed in the transfer cavity so that the mold is filled and pressure maintained. After an interval to allow the cure to be completed, the pressure is released and the finished molding and excess charge in the transfer cavity is ejected.

The advantages of transfer molding are that the flow ensures greater homogeneity of the material and that the molding dimensions are independent of charge weight, provided that there is a material excess.

Reaction Injection Molding (RIM)

This is a low-pressure process, essentially cold curing, although the molds can be heated to ensure a good surface finish on the molding. The process is applicable principally to polyurethane-based systems where two liquid precursors, a polyol and a diisocyanate, are

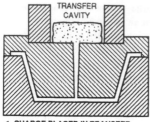

1. CHARGE PLACED IN TRANSFER CAVITY

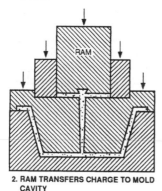

2. RAM TRANSFERS CHARGE TO MOLD CAVITY

FIGURE 3.4-6.

mixed in metered proportions just as they are injected into the mold cavity (Figure 3.4-7). The plant is relatively complex. Apart from the mold opening and closing mechanism, a series of pumps and accumulators is required to prepare the charge. Predetermined metered flows of the two ingredients must be delivered via separate pipes to the mold. The runner system in the mold is designed so that the two liquid streams impinge and mix in a series of turbulent mixing chambers before flowing into the mold cavity. In a typical system the

REACTION INJECTION MOLDING (RIM)

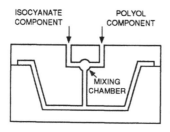

FIGURE 3.4-7.

liquids are each pumped into a hydraulic accumulator until a sufficient charge at the correct pressure is obtained. A series of electromagnetic valves then directs the flow of liquids to the mold. There is a recirculating system to deal with any excess.

The key to the process is that the component liquids never mix except in the mold. Once mixed, they cure rapidly at ambient temperature. The cured material can be rigid, rubbery, or a foam, according to the formulation of the precursors. If the resin is to be reinforced, then chopped fibers have to be blended into (usually) both precursors prior to injection. Alternatively, reinforcement may be pre-positioned in the mold before injection (R-RIM). This process is a promising candidate for mass production of advanced composite structures and parts.

Injection Molding

This is a high-pressure process principally applicable to thermoplastics-based systems, although variants of the process are used for thermoset (and even BMC) compounds.

For thermoplastics, the feedstock is usually precompounded molding pellets which contain 10–40% by weight of very short glass (or graphite) fibers. These are hopper-fed to the plasticization stage of the machine which is now invariably a screw-pump device. This device conveys the charge down a heated barrel while subjecting it to a strong shearing action that results in a viscous homogeneous mix. The charge is normally carried to a cavity in front of the screw, which is closed by a cutoff valve. The screw moves backwards as the pressure of charge builds up in the cavity until a sufficient charge has accumulated. When the rotation of the screw is stopped, and the whole screw is forced forward by a hydraulic ram, the cutoff valve is opened, and the charge injected under a very high pressure, typically $1.00–2.00 \times 10^7$ kg/m² (15,000–30,000 psi), into the mold cavity. The mold is kept relatively cold and is held closed, usually by another hydraulic system. When the molding has solidified, the pressure is relaxed while the next charge is prepared. As soon as the molding is cool enough to be handled without distorting it, the mold is opened and the part ejected; the process is then repeated.

The injection molding process is capable of producing intricate parts, from a few grams weight to over 100 kg (220 lbs), to very accurately controlled dimensions.

The cycle time can be as little as six seconds for very small components up to a few minutes for the largest ones. Small articles can be produced at very high rates in multicavity molds. The cost of the mold is high, and the general capital outlay for the molding machine is also high when compared with the alternative fabrication processes discussed. This process can seldom be justified unless more than 10,000 parts are required. A single cavity mold operated on a 20 sec cycle is capable of producing upwards of one million (1 M) parts in a working year. There is an increasing trend towards the use of this process to make large intricate moldings, e.g., truck dashboards. These parts require very large molding machines with mold-closing forces upwards of 2,500 tons. The process is, of course, very widely used with unreinforced thermoplastics, and, at present, the reinforced grades account for less than 10% of the total. Reinforced grades are more difficult to mold because of the increased viscosity and other rheological changes produced by the fiber addition [5]. The fibers are also abrasive, which can lead to rapid wear in both machine and mold unless appropriately hardened metals are employed.

The injection molding of thermosets is relatively straightforward. The only differences are that the mold is now heated to a relatively high temperature and the thermoset must be transferred to the mold at a low temperature so that it does not gel too rapidly. Fast injection through a small gate is often used to introduce adiabatic heating to the charge as it enters the mold; fast injection makes for a more rapid cure and reduces the total cycle time. The process is also being adapted for DMC/BMC materials. Here the barrel can be kept quite cool, as the charge is liquid even at room temperature. The fiber length in DMC/BMC is generally much greater than in conventional injection molded thermoplastics, and, for this reason, only the larger sizes of machines are likely to be capable of handling the material. The screw design must be modified to suit the flow characteristics of the charge, which is a relatively new development, and processing guidelines have yet to be established.

Press Molding

Press molding is used for making flat, or slightly curved, panels from prepreg and sometimes from wet impregnated sheet reinforcements. The essentials of the process are that the laminate is compressed between a pair of flat, or matching, curved platens which are heated. A relatively low pressure is applied, and in many cases, the thickness of the laminate is controlled by pressing the platens to fixed stops. Provision is made for excess resin to escape so that a laminate of correct fiber content can be produced. In the case of prepreg, it is common to cover the pack of prepregs with a sheet of perforated or porous release film and to place bleeder cloths (glass cloth or sometimes absorbent paper) on the outside. These cloths absorb the excess resin. This procedure is clearly an expensive one justified only when high-performance moldings are being produced. In this type of operation, it is often critical to apply the pressure at the moment of resin gelation—neither earlier nor later. The former action would result in an excessive resin flow and loss, whereas the latter would not allow the resin to flow sufficiently. These remarks apply principally to epoxy resin composites.

Press molding is also used to produce flat and corrugated sheet in glass/polyester. In this case the conditions are usually optimized to produce a consistent clarity and surface finish, the mechanical properties of the molding being less demanding. A somewhat similar forming process, "pressure molding," is being developed for thermoplastic composites [6].

Autoclave Molding

This process is widely used in the aerospace industry for the production of high quality flat or curved panels from prepreg material. The laminate is built up onto a mold plate, usually of metal, conforming to the shape of the panel to be produced. It is usual to cover both sides of the pack with a single layer of a fine polyester cloth known as a "peel-ply." (This cloth is stripped off after molding, leaving a clean, smooth surface to the laminate.) The top surface of the laminate is covered with a perforated, or porous, release film and, if necessary, "bleeder" or "breather" cloths; the whole assembly is then covered with a non-porous membrane which is sealed to the molding plate. Finally, the whole assembly is loaded into the autoclave (Figure 3.4-8). This method of molding allows a vacuum to be drawn between the molding plate and the cover membrane while the pressure and temperature within the autoclave are separately controlled [7].

The vacuum serves to continuously remove all volatiles that might form during the molding operation to reduce the incidence of porosity. The combined effects of the vacuum and autoclave pressure acting on the cover membrane serve to apply a very even pressure over the entire surface of the molding. This technique offers a much greater delicacy of control than is available in alternative molding techniques, especially when only low molding pressures, e.g. $1–10 \times 10^4$ kg/m² (15–150 psi), are required. In many prepreg systems, the most critical processing requirement is to apply the molding pressure when the resin is at the correct state of cure (i.e., correct viscosity). In many systems, when the resin is first heated to the curing temperature, it melts to a very low viscosity liquid which ensures good wetting and bonding between the prepreg layers. The application of two much pressure at this stage can result in excessive bleeding and a "dry" laminate with too low a resin content. Conversely, too little pressure or late application of pressure (past gelation) can lead to high resin content and a porous laminate. The gel point is usually established by checking the gel time for each batch of prepreg before starting molding operations. It is also possible to monitor the "state of cure" during the actual molding operation by dielectric loss measurements [8], for instance. In some systems, where a very mobile resin dictates very low molding pressures, the autoclave itself is also evacuated, but a vacuum differential between the molding bag and the autoclave interior is maintained to achieve the desired pressure.

The current trend in prepreg technology is towards modified resins with higher melt viscosity and to the adoption of a zero-bleed technique. In these cases, no excess resin is incorporated in the prepreg, and the high resin viscosity prevents significant bleeding through the porous release membrane.

The autoclave process is extremely versatile but, generally, rather slow, and the capital cost of the plant is relatively high. The advantage of the process is that consistent moldings of very high quality can be produced. These features combine to make the process attractive in the aerospace and high technology areas, but less so in the mass production, consumer durables, or automotive industries.

Pultrusion

This process involves drawing the reinforcement through a bath of liquid resin and then directly and continuously through a heated die to produce a continuous section. The process is most easily operated when unidirectional reinforcement is required. In this case, a sufficient number of rovings, or tows, of fiber are combined to produce the required fiber volume fraction across the section to be produced. It is possible to incorporate a proportion of off-axis reinforcement by drawing woven tapes and/or strips of chopped strand mat together with the continuous rovings.

The process is suitable for simple sections such as circular rods, tubes, channel, and I-sections. Very good fiber alignment and very high fiber fractions can be obtained, which leads to excellent mechanical properties.

The critical parameters of the process are the drawing velocity and the mold temperature [9]. The mold length is determined by the cure characteristics of the resin. Ideally, a fast-curing resin with low exotherm and low volatile content is preferred. Draw speeds are generally quite low, of the order of 1–10 m/min (3–30 ft/min). Pultrusions may be made from glass or graph-

AUTOCLAVE MOLDING

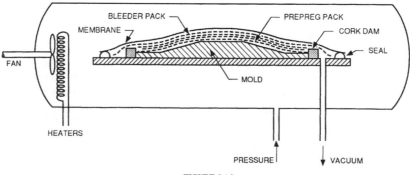

FIGURE 3.4-8.

Table 3.4-1. Fiber reinforcement.

Materials	Application	Availability and Cost Index*	Materials	Application	Availability and Cost Index*
E-Glass			*Kevlar*		
Roving	Filament Winding Pultrusion SMC/BMC/DMC Manufacture Spray-Up Molding Prepreg Manufacture	A.1	Roving/Yarn	Filament Winding Pultrusion Prepreg Manufacture	C.4
			Cloth	Contact Molding Matched-Die Molding Autoclave Molding	C.4
CSM (chopped-strand mat)	Contact Molding Resin Injection Molding Matched-Die Molding	A.1	*High-Strength Graphite*		
Woven Roving	Contact Molding Resin Injection Molding Matched-Die Molding	A.2	Tow	Filament Winding Pultrusion Prepreg Manufacture	C.4
Woven Cloth	Contact Molding Resin Injection Molding Matched-Die Molding	A.2	Cloth	Contact Molding Autoclave Molding	C.4
Chopped Roving	BMC/DMC Manufacture FR Thermosets FR Thermoplastics	B.2	Chopped Fiber Mat	Contact Molding	D.4
			Chopped Fiber	FR Thermoplastics FR Thermosets	C.4
S-Glass			*High-Modulus Graphite*		
			Tow	Filament Winding Pultrusion Prepreg Manufacture	C.5
Roving	Filament Winding Pultrusion Prepreg Manufacture	C.3	*Boron*		
Woven Roving	Contact Molding	D.3	Continuous Fiber	Filament Winding Prepreg Manufacture Metal Matrix Systems	C.5
Woven Cloth	Autoclave Molding	D.3			

*See p. 100 for definitions.

ite fibers, and both polyester and epoxy resin systems are widely used. Pultrusion of thermoplastic matrix composites is currently under development [10].

3.4.2 Summary of Economic Factors for Basic Systems

In the foregoing discussion we have seen that processing of fiber reinforced composite materials can take a great variety of forms. In this subsection the relevance of these materials and processes to manufacturing requirements will be examined.

In making the decision to produce an article in a fiber reinforced material, the overriding considerations must be performance and cost-effectiveness. In aerospace applications, performance is frequently more important than cost. But for general industrial applications, cost will generally be the final deciding factor, except for those applications where weight-saving is an overruling consideration. As a general rule for non-aerospace applications, the constituents of the composites, e.g., glass fiber and resin, are more costly on a weight basis than the traditional materials we seek to replace with them.

How then can composites be cost-effective? The answers are complex. The composites are low-density materials with highly specific mechanical properties. Thus, on a volume basis or on a cost per unit stiffness or strength basis, they can be more competitive. Also, as noted above, low density in itself is becoming increasingly important, especially in the automotive industry. Composites are amenable, in many cases, to being fabricated into complex shapes so that a single

Table 3.4-2. Resin materials.

Material	Application	Availability and Cost Index*	Material	Application	Availability and Cost Index*
Unsaturated Polyester			*Phenolic**		
Orthophthalic Based	General Purpose Resin Contact Molding Spray-up Matched-Die Molding	A.1		Thermosetting Resins Compression and Transfer Molding Injection Molding Hot Press Lamination	A.2
Isophthalic and Terephthalic Based	Premium Quality Resins Contact Molding Spray-up Resin Injection Molding Matched-Die Molding SMC/BMC/DMC Filament Winding Pultrusion	A.2	*Polyimide**	High Temperature Resistant Thermoset/Thermoplastic Prepreg Press and Autoclave Molding	D.5
Bisphenol A Based	High Quality Corrosion Resistant All Molding Processes	B.3	*Thermoplastics**		
Alkyd and Glyptal Resins	Hot Curing Systems BMC/DMC Thermoset Systems	A.2	Polyamide (Nylon)	High-Strength Thermoplastic Injection Molding	A.3
Epoxy Resins			Polypropylene	General Purpose Thermoplastic Injection Molding	A.1
Bisphenol A Based	General Purpose Resins Contact and Die Molding Filament Winding Prepreg Pultrusion	B.3	Polycarbonate	High Impact Strength T/P Injection Molding	A.3
			A.B.S.	General Purpose T/P Injection Molding	A.2
Cyclo-aliphatic and Poly-functional Resins	High Heat Distortion Resins All Fabrication Processes	C.4	Polyether Sulphone	High Temperature Resistant T/P Injection Molding	C.4
Vinyl Ester			Polyphenylene-sulphide (PPS)	High Temperature Resistant T/P Injection Molding	C.4
	Premium Quality Resins All Contact Molding Matched-Die Molding Filament Molding Pultrusion SMC	B.3	Polyether-ether Ketone (PEEK)	High Temperature Resistant T/P Injection Molding Most Manufacturing Processes	B.4

*Comments apply *only* to fiber reinforced grades.

Table 3.4-3. Fiber reinforced plastics fabrication.

Process	Applicable Materials			Molding Conditions			Production Economics		
	Resins	Glass Fiber	Other Fiber	Temperature	Time	Pressure	Labor Content	Capital Requirement	High Production Rate Capability
Contact Molding Hand Lay-up	UPE VE EP	CSM C/WR	GR K	A	M/L	O/V	H	L	X
Contact Molding Spray-up	UPE VE	CR	(K)	A	M/L	O/V	M	M	✔
Matched-Die Molding (COLD)	UPE VE	CSM C/WR	K (GR)	A	S/M	L	M	M	✔
Matched-Die Molding (HOT)	UPE VE	CSM C/WR	K (GR)	M	S/M	M	M	M	*
	EP	CSM C/WR	K GR	M/H	S/M	M	M	M	*
SMC (Hot Die) BMC DMC	UPE VE	SMC BMC DMC	K	M/H	S	M/H	L	H	*
	EP	SMC		H	S	M/H	L	H	*
Resin Injection Molding (Resin Transfer Molding)	UPE VE	CSM C/WR	K (GR)	A	M/L	L/V	M	M	✔
	EP	CSM C/WR	K GR	A/M	M/L	L/V	M	M	✔
Reaction Injection Molding (RIM, RRIM) ('Prepositioned)	PU	CH CSM' C/WR'		A	S	M	L	M	*
Hot Pressing	UPE VE EP	CSM C/WR PP	K GR	M/H	S/M	M	M	M	*
Autoclave Molding	(UPE) VE EP	(CSM) C/WR PP	K GR	M/H	M	L/V	M	H	X
Filament Winding	UPE VE EP	CR C/WR PP	K GR	A/M	M	O	L	H	✔
Pultrusion	UPE VE EP	CR	K GR	M/H	S/M	O	L	H	*
Compression Molding	UPE VE EP	DMC BMC		M/H	S	H	L	M	*
	TS	CH/M		H	S	H	L	M	*
Injection Molding	TP	CH/M	GR	M/H	S	VH	L	H	*
	TS	CH/M		H	S	H	L	H	*
	UPE VE	BMC DMC		M	S	H	L	H	*

Table 3.4-4. Contact molding resins.

Description	Viscosity at 25°C (NSM^{-2})	Heat Distortion Temperature °C	Remarks
General Purpose Orthophthalic P. Ester	0.9	72	
General Purpose Isophthalic P. Ester	0.6	130	High Heat Distortion Temperature
Isophthalic Resin for Rapid Impregnation	0.4	80	Low Viscosity, Low Exotherm
Bisphenol A Based Resin	0.8	120	High Heat Distortion, Good Chemical Resistance
Orthophthalic Gel Coat Resin	0.9	60	Thixotropic
Isophthalic Gel Coat Resin	1.2	80	Thixotropic, High Water Resistance
High Impact Vinyl Ester	0.6	133	High Heat Distortion, Stiffer and Tougher Than Polyester (Hot Cured)
High Temperature Vinyl Ester	0.6	150	V. High Heat Distortion (Hot Cured

molding can replace a whole assembly of parts. This "parts consolidation" effect also brings savings in labor, inspection, storage, and packaging. Additional savings can result from secondary properties. For instance, excellent corrosion resistance of SMC moldings eliminates the need for expensive under-body protection schemes in automobiles. Other important properties are electrical insulation, thermal insulation, and "self-finish," i.e., no requirement for further decoration.

Another very important factor favoring the adoption of composites is the availability of a processing route suited to the application. For example, for a very low volume application such as racing car bodies, a hand lay-up contact molding in glass/polyester may be highly cost-effective, though not suitable for mass production. For large volume automobile components, where cost dictates automated systems, such techniques as SMC, injection molding, and reaction injection molding (RIM and R-RIM) are better suited.

In the tables which follow, materials and processes are compared on the basis of their overall applications potential. In Tables 3.4-1 and 3.4-2, resins and fibers are rated on a simple availability and cost index, whereas in Table 3.4-3, the materials applicable to each listed process are indicated together with comparative processing conditions and production potential. Supplemental information on contact molding resins is included in Table 3.4-4.

Availability and Cost Index (Tables 3.4-1 and 3.4-2):

- Availability
  - A   Tonnage quantities readily available
  - B   Readily available in moderate quantity
  - C   Restricted supply
  - D   Very limited supply
- Cost
  - 1   Lowest cost in class
  - 2   Premium over base material
  - 3   Moderately expensive
  - 4   Expensive
  - 5   Very expensive

In Table 3.4-3, the following abbreviations are used:

- Column
  - a) Resins
    - UPE   Unsaturated polyester
    - VE   Vinyl ester
    - EP   Epoxy
    - PU   Polyurethane
    - TS   Thermoset
    - TP   Thermoplastic
  - b) Glass Fiber
    - CR   Continuous roving
    - CSM   Chopped strand mat
    - C   Cloth
    - WR   Woven roving
    - CH   Chopped fiber
    - M   Milled fiber
    - SMC   Sheet molding compound
    - BMC   Bulk molding compound
    - DMC   Dough molding compound
    - PP   Prepreg
  - c) Other Fibers
    - K   Kevlar® Polyaramid
    - GR   Graphite
  - d) Temperature
    - A   Ambient—20–30°C

M   Moderate—50–150°C
H   High—over 150°C
e) Time
   S   Short—under 10 min
   M   Moderate—10 min–1 hr
   L   Long—over 1 hr
f) Pressure
   O   No pressure
   L   Low—less than 10 bar (150 psi)
   M   Moderate—10–100 bar (150–1500 psi)
   H   High—over 100 bar (1500 psi)
g) Labor Content
   L   Low
   M   Moderate
   H   High (labor-intensive)
h) Capital Requirement
   L   Low
   M   Moderate
   H   High
i) High Production Rate Capability
   X   Not suitable for high-production rate
   ✔   Suitable
   *   Very high rates possible

NOTE: A circled symbol, e.g., (GR), indicates not normally recommended.

An underlined symbol, e.g., UPE, indicates the combination most commonly selected for the particular process.

The information in the tables is intended to be used as a first basis for selection. Only the principal variations are indicated, and they are not intended to embrace all possibilities.

3.4.3 References

1. MORENA, J. J. *Advanced Composite Mold Making*. Van Nostrand Reinhold Co., New York (1988).
2. COULTER, J. P. and S. I. Güçeri. "Resin Impregnation During the Manufacture of Composite Materials Subject to Prescribed Injection Rate," *J. Reinforced Plastics and Composites*, 7:200 (1988).
3. GAUVIN, R., M. Chibani and P. Lafontaine. "The Modeling of Pressure Distribution in Resin Transfer Molding," *J. Reinforced Plastics and Composites*, 6:367 (1987).
4. BARONE, M. R. and D. A. Caulte. "A Model for the Flow of a Chopped Fiber Reinforced Compound in Compression Molding," *J. Applied Mechanics*, 53:361 (1986).
5. GIBSON, A. G. and A. N. McClelland. "Rheology and Packing Effects in Injection Molding of Long Fiber Reinforced Composites," *Proc. ICCM & ECCM, Vol. 1*. F. L. Matthews et al., eds., Elsevier Appl. Sci., London, p. 1.131 (1987).
6. WITZLER, S. "Pressure Molding Methods for Thermoplastic Composites," *Advanced Composites*, 3(5):49 (1988).
7. CARLSSON, L. A. and R. B. Pipes. *Experimental Characterization of Advanced Composite Materials*. Prentice-Hall, Englewood Cliffs, NJ (1987).
8. DAY, D. R. "Effects of Stoichiometric Mixing Ratio on Epoxy Cure—A Dielectric Analysis," *J. Reinforced Plastics and Composites*, 7:475 (1988).
9. MA, C. C. M., J. S. Hwang and W. C. Shih. "Effects of the Processing Parameters on Pultrusion Process," *Proc. ICCM & ECCM, Vol. 1*. F. L. Matthews et al., eds., Elsevier Appl. Sci., London, p. 1.110 (1987).
10. NOLET, S. C. and J. P. Fanucci. "Pultrusion and Pull-Forming of Advanced Composites," *Proceedings from a Workshop on Design, Manufacture and Quality Assurance of Low Cost, Lightweight Advanced Composite Molds, Tools and Structures*, Florida Atlantic University, Boca Raton, FL (October 17–19, 1988)

Open Mold Laminations— Contact Molding

3.5 OPEN MOLD LAMINATIONS — CONTACT MOLDING

M. G. BADER

3.5.1 Introduction

Open mold, or contact lamination, is the oldest of the methods employed for fabricating fiber reinforced laminates. In its basic form, it is a versatile, but slow and labor-intensive, process. Although quality control rests almost entirely in the skill of the operator, since the process is fundamentally very simple, operators of low skill are often employed. Thus, in the past, the quality of laminates has tended to be very variable. A number of developments of the process have been introduced with the aim of reducing the labor content, increasing the production rate, and assuring a consistent quality in the product. These developments are discussed later, but first we will discuss the basis of the process and the variables that control the properties and quality of the product.

The essentials of the process have already been outlined in section 3.4 (Figure 3.4-1). The laminate is formed by laying down a number of layers of reinforcement which may be CSM, woven roving, cloth, or continuous random mat. These layers must be uniformly impregnated with liquid resin and consolidated to form a laminate, free from voids, of the desired thickness and with the correct reinforcement to resin ratio.

The mechanical and physical properties of the laminate are determined by the choice of resin and reinforcement, the reinforcement to resin ratio, and the void content. The degree of penetration of the liquid resin into the fiber bundles and the nature of the fiber/resin interface are also important. The chemical and thermal resistances of the laminate are most strongly influenced by the choice of resin and the conditions under which it is cured (see section 3.3).

Since contact molding is such a versatile process, embracing moldings ranging from less than 1 kg (2.2 lb) to ship hulls of 85 m (280 ft) in length, there must be many variations in the detail of the operation and the materials utilized. The most important combination, however, is that of E-glass fiber with unsaturated polyester resin. The resins are to some extent now being replaced by vinyl ester resins, and some applications use graphite or Kevlar fiber alone or in combination with glass. Epoxy resins are also used to a limited extent. However, the use of these more expensive materials tends to be restricted to more demanding applications and low volumes.

An important point concerns health hazards. The unsaturated polyester resins contain up to 40% of styrene monomers, which are responsible for their characteristic odor. Styrene is considered hazardous in the atmosphere; in general, levels must be kept below 5–10 ppm in working areas. This level is quite difficult to achieve when large hand-laid moldings are made; it calls for sophisticated ventilation equipment. Volatile amines and solvents in other resins also constitute potential health hazards. Most liquid resins are also highly flammable, and peroxide promoters (catalysts) can constitute an explosion hazard if mishandled. In the following sections the influence of the major variables on the process and product are considered together with the variations on the basic process.

3.5.2 The Mold

The mold may be of either male or female form. The latter form is more common, but the choice is dictated by which surface of the molding is required to have the prime finish, since only the "contact" surface of the molding will be "fair." The mold material may be plaster, concrete, wood, GRP, or metal [1]. For medium production series, GRP is preferred; for long runs, metal (usually steel) is used. For "one offs" and prototypes, plaster molds are commonly used. The ultimate choice is dictated by the size of the molding, the length of the production run, the degree of perfection required on the prime surface, and whether heat assistance, vacuum bags, or both are to be used during the cure cycle.

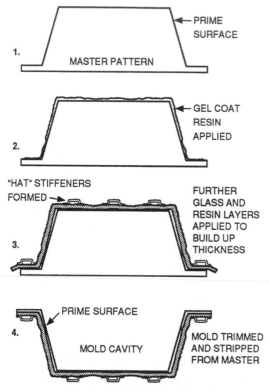

1. MASTER PATTERN
PRIME SURFACE

2. GEL COAT RESIN APPLIED

3. "HAT" STIFFENERS FORMED
FURTHER GLASS AND RESIN LAYERS APPLIED TO BUILD UP THICKNESS

4. PRIME SURFACE
MOLD CAVITY
MOLD TRIMMED AND STRIPPED FROM MASTER

FIGURE 3.5-1. Principle of GRP mold production.

A typical medium-sized molding, e.g., a boat hull or bath tub, would utilize a GRP mold. This mold is made from a master form, or "plug," which must be finished to the correct dimensions and surface finish required of the production molding. The mold, which might be of several pieces, is made by laminating GRP onto the master after having the surface polished and treated with a suitable release agent (polyvinyl alcohol emulsion, various waxes, or silicone compounds). The first layer applied to the master is usually a lightly filled resin. Polyesters are often used, but epoxies are usually preferred when medium to long production runs are envisaged. The initial surface coating is then backed up by fine glass surfacing tissue with more resin and, subsequently, further layers of reinforced resin, until the mold has the requisite thickness and rigidity for the mold dimensions and for the molding process to be used. A typical mold construction sequence is illustrated in Figure 3.5-1. Stiffening sections are usually molded onto the back of the mold, and mating flanges, etc., are also constructed in the case of multipart molds. Larger molds will incorporate frameworks and jigging points to facilitate handling on the production line. With care, several hundred moldings may be made from a GRP mold, but when very long runs are envisaged, multiple molds will be required. All these molds are taken from the original master, which must be of suitably durable construction according to the number of molds envisaged. Where only a few molds are required, the master will be constructed from wood, usually finished with an epoxy resin coating. Smaller moldings may use plaster masters, whereas for really long runs where many molds are required, a metal master may be called for. Likewise, when the production run is very long, GRP molds are often replaced by metal molds. These metal molds are not, of course, produced from masters but must be fabricated or cast and machined to form. They are very costly, but durable. An additional advantage of metal molds is their high thermal conductivity, which can be useful in dissipating the heat generated during the cure exotherm. Metal molds can also be heated if a heat-assisted cure is specified. Cast metal molds may be made from aluminum or zinc alloys, and the mold cavity can be chromium-plated or hard-surfaced and polished to produce a durable surface with excellent molding finish. Such molds are convenient for medium-sized moldings up to about 2 m (6 ft) maximum dimension. Above this size they become relatively expensive, so that GRP or sheet steel would be preferred. The very largest molds for ships are generally fabricated from welded steel plate, suitably stiffened.

The choice of mold material is, thus, a function of the size of the molding and the length of the production run. Likewise, the number of molds required will be determined by the production rate, the lay-up time, and the cure time. The cure time is generally the controlling parameter.

3.5.3 Choice of Reinforcement

Glass Fiber

The two basic alternatives in open mold contact molding are "hand lay-up" and "spray-up." The former one requires the reinforcement to be in the form of sheets or a web, whereas in the latter case, chopped fiber rovings are sprayed onto the mold [2]. Elements of both processes can be combined.

For hand lay-up operations, chopped strand mat (CSM), continuous random mat (CRM), cloths, and various woven roving constructions are available. The choice is governed by economics and by the properties (notably the mechanical properties) required of the part. CSM provides reinforcement giving mechanical isotropy in the plane of lamination, but the maximum glass content is limited to about 40% by volume, with 35% being a more typical value. Cloths, on the other hand, give an orthotropic, biaxial reinforcement in the plane of lamination, and their more regular construction allows higher glass contents, hence, the achievement of larger stiffness and strength. The heavier weight woven rovings allow even higher glass contents, and unbalanced or unidirectional forms are available (see Figure 3.2-1) which allow the properties to be maximized in one direction. The highest stiffness and strength (in one direction) are achieved with parallel plies of unidirectional woven roving. If biaxial properties are required, laminates of unidirectional woven rovings with the alignment rotated 90° between adjacent plies give marginally better properties than laminates of balanced woven rovings or cloths. CSM and some cloth weaves, such as "twill" or "satin," drape more readily than woven rovings and, thus, conform more readily to complex mold contours. The relative mechanical properties of laminates fabricated from CSM, cloth, and woven roving are shown in Figures 3.5-2a and 3.5-2b (see also reference [3]).

Another important consideration is the weight of the mat or cloth used. Clearly, the design properties will require that there be a certain amount of glass fiber in the molding. This requirement may be achieved with fewer layers of a heavy weight reinforcement or more of a lighter material. The lighter material, however, will usually be more convenient to handle, will wet-out more rapidly, and will give a finer textured surface (on the back face).

In recent years, fiber manufacturers have devoted considerable effort towards developing glass-fiber finishes that facilitate wetting of the fibers by the liquid resin while conferring the desired stability to the web. The fast-wetting grades of CSM are now very widely used and can reduce the lay-up time by as much as 30% in comparison with earlier products.

To summarize, when the very highest stiffness and strength are required, the molding should incorporate unidirectional woven rovings laid in a sequence to provide the correct balance of properties. Balanced woven rovings give a biaxial distribution of properties in the laminating plane only slightly inferior to alternating 0° and 90° unidirectional material. Fine cloths drape well and give good surface finishes, whereas CSM and CRM give isotropic in-plane properties, albeit at a lower level. From the economic viewpoint, CSM is generally the least expensive form of reinforcement and also the easiest and fastest to fabricate, so that general purpose moldings requiring only moderate mechanical properties will generally use only this material. For more demanding applications, some woven roving will be incorporated. A particularly effective combination consists of alternate layers of woven roving and CSM. This combination gives good stiffness and strength and, possibly, even better interlaminar shear strength than all-woven roving constructions.

The spray-up process uses an in situ chopped roving. The laminate is essentially random, and the only variables are the strand and fiber count of the roving. Generally, the heaviest roving consistent with ease of handling and chopping will be used—typically 2,400 tex (207 yards/lb).

The chemical and water resistance of the molded part is not strongly influenced by the form or quantity of the reinforcement but is strongly dependent on the resin properties, the degree of cure, and the correct choice of coupling agent incorporated in the fiber finish. In general, silane or "chrome-silane" finishes give the best properties with polyester resins. It is also important that the fiber bundles are completely wetted by the resin and that the laminate is free from dry spots, voids, and microporosity which might serve to allow water or a more harmful contaminant to infiltrate the molding. These factors are all influenced strongly by the fiber and mat finish and also are dependent on the fabrication procedure. The correct choice of reinforcement will allow a fully wetted, high quality laminate to be produced in the shortest possible time.

Graphite and Kevlar Fibers

Graphite, Kevlar, and (to some extent) S-glass fibers are alternatives to E-glass for reinforcement of open mold products. They offer higher specific stiffness than E-glass but at a considerably higher cost. Since these fibers are expensive, they are seldom used in the form

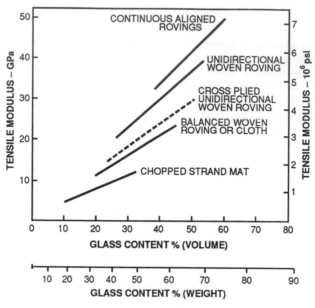

FIGURE 3.5-2a. Stiffness of glass/polyester laminates showing effect of different forms of reinforcement over practical range of glass content.

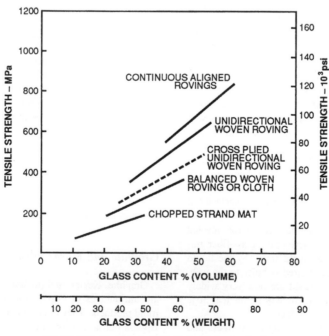

FIGURE 3.5-2b. Tensile strength of glass/polyester laminates.

of random mat, but rather in cloth, woven roving, or prepreg form. The latter is not applicable to open mold operations. One of the more cost-effective ways of using graphite and Kevlar fibers is to obtain localized stiffening of GRP moldings or to use them as hybrids containing a relatively small proportion of the high-grade fiber. Kevlar is available as cloth and woven roving of various weights; graphite is available as woven roving and also as a hybrid woven roving with glass. In this latter form some of the glass rovings in a woven roving product are substituted by graphite. This is a convenient manner in which to apply a small proportion of graphite to a GRP molding. In addition to its higher stiffness, Kevlar also enhances toughness and, in some circumstances, corrosion resistance. Graphite, being an electrical conductor, is sometimes used for electromagnetic screening applications on the surface of GRP.

3.5.4 Choice of Resin

The primary choices for contact molding processes are the three main categories of laminating resins: the unsaturated polyesters, the vinyl esters, and the epoxies. For contact molding, epoxies are seldom used other than for short runs of specialized moldings and for moldings requiring specific chemical or elevated temperature resistance. The discussion in this subsection will therefore be confined to the polyester and vinyl ester materials, with the emphasis on the former one.

Among the factors influencing this choice are the following:

a. The properties required of the laminates that are influenced by resin
   1. Mechanical: stiffness, strength, and toughness
   2. Physical: density, electrical resistivity, and dielectric loss
   3. Thermal: heat distortion temperature, glass transition temperature, and thermal degradation characteristics
   4. Chemical: resistance to water, acids, alkalis, etc.
b. Fabrication characteristics
   1. Storage life of components before and after compounding
   2. Physical properties of liquid resin: viscosity, thixotropy, wetting characteristics, and ease of molding
   3. Cure characteristics: cure temperature, cure rate, exotherm, shrinkage, and volatiles evolved

4. Safety: environmental hazards due to inhalation or contact, fire, and explosion risks
c. Economics
   1. Cost of constituents and fabrication in relation to the properties and production rate achieved—cost-effectiveness

Unsaturated Polyester Resins

Unsaturated polyester resins are extremely complex from a chemical viewpoint (see section 3.3). The base resin is formed from three principal classes of constituents:

1. Unsaturated acids (or anhydrides)
2. Saturated acids (or anhydrides)
3. Glycols

These constituents are reacted to form polyesters of a moderate degree of polymerization. The most influential factors on the properties of the final resin are the proportion of unsaturated acid and the degree of polymerization. The unsaturated acid provides the sites for cross-linking via the styrene bridge during final cure, and a higher proportion of the unsaturated component will result in a higher cross-link density and a harder, more brittle, and more chemically resistant resin. However, the choice of saturated acids and glycols is also important. Aromatic acids give harder, more rigid, heat resistant, and chemically resistant products. Conversely, the higher glycols produce a more flexible backbone structure.

The other principal constituent of the resin is monomeric styrene, or an alternative, vinyl monomer. This constituent is added in amounts of up to 50% to the final resin. The function of the styrene is to respond to the promoter and to form cross-links between unsaturated sites in the polyester molecules by means of polystyrene bridges.

In addition to these principal ingredients, inhibitors, diluents, plasticizers, and, of course, the promoter and accelerator are also added to the base resin. Their interactions are complex and the alternative formulations virtually infinite. However, in practice, the basic parameters can be summarized as follows:

a. Choice of saturated and unsaturated acids
b. The unsaturation ratio—unsaturated acid/saturated acid
c. The choice of glycol

d. The degree of polymerization of the base polyester
e. The resin:styrene ratio
f. Choice of promoter and accelerator, and inhibitor

These are discussed briefly in the following paragraphs.

a. Acid Components. The unsaturated acid is almost always maleic acid (anhydride) or its isomer, fumaric acid. The latter co-polymerizes more readily with styrene. A greater proportion of fumaric acid will give a faster cure and a higher exotherm for a given level of promotion. Fumarates are claimed to give longer shelf life than maleates in the uncured resin. They also form resins with higher viscosity and exhibit greater shrinkage on curing. It should be noted that the acid is almost always added as maleic, but a proportion of it becomes converted to fumaric during the esterification reaction. It follows that mixes that are reacted for longer periods (higher degree of polymerization) tend to have higher fumarate unsaturation.

The choices of saturated acid are several. The principal contenders are the aliphatic, malonic, adipic, and sebacic acids; the three isomers of the aromatic phthalic acid (ortho-, iso- and tere-); certain halogenated derivatives of these acids—notably the so-called "HET-acid" (hexachloro-endomethylene-tetrahydrophthalic acid).

The aliphatic acids give flexible resins with low softening temperatures and poor chemical resistance. A small proportion of these acids may be incorporated to increase flexibility.

The phthalic acids are virtually universally used in commercial resins. Orthophthalic is the least expensive and is the basis of many commercial resins. It forms resins with softening temperatures around 70°C and moderate chemical resistance. Isophthalic acid gives higher softening temperatures (typically 90°C) and much better chemical resistance. It is the basis of many "premium grade" resins. Terephthalic acid is more expensive than isophthalic but allows even higher softening temperatures, approaching 200°C in some cases, and gives superior chemical resistance. Halogen-substituted acids (e.g., "HET-acid") are used in the formulation of fire resistant resins.

b. Unsaturation Ratio. A higher ratio of the unsaturated acid gives a resin that cures to a higher cross-link density and that is more rigid, brittle, and higher in heat distortion temperature. A high unsaturation ratio also leads to greater exotherm, shorter shelf life of the uncured resin, and greater shrinkage on cure. A reasonably practical compromise lies in the range of US:S from 1:1 to 2:1. Low shrinkage, low exotherm resins will tend towards lower unsaturation levels. The unsaturated acids are generally more expensive than ortho- and isophthalic acids.

c. Glycol (Diol). The properties of the cured resin are as strongly influenced by the choice of the glycol (diol) as of the acid. Although the potential choice is enormous, economic factors dictate only a few alternatives. The aliphatic glycols form a series from ethylene glycol through propylene glycol to higher (longer chain) glycols. In general, the higher glycols correspond to more flexible and less chemically resistant cured resins. Thus, ethylene glycol gives the highest stiffness and heat distortion temperature of the aliphatic glycol series. However, propylene glycol is less expensive and gives only marginally inferior properties which, in practice, may be no detriment, depending upon the flexibility specified in a given formulation. Thus, propylene glycol is the most commonly utilized glycol in commercial formulations.

Stiffer, more chemically resistant resins can be produced if aromatic glycols are utilized, either alone or in combination with aliphatic glycols. The most popular glycol is the derivative of bisphenol A, known as hydrogenated bisphenol A (HBPA). Its use gives rise to the so-called bisphenol A derived polyesters (see section 3.3). These resins appear to have the best chemical and water resistance of all the unsaturated polyester resins.

d. Degree of Polymerization. This is controlled by the temperature and time at which the esterification reaction is allowed to proceed. The longer the reaction, the higher the molecular weight and the viscosity of the resin. The practical measure of the degree of cure is the proportion of unreacted acid remaining in the mix, the well-known "Acid Value," which is, in fact, the number of milligrams of potassium hydroxide required to neutralize one gram of the resin. Acid values in the range 25–50 cover most commercial resins. A lower value indicates a higher degree of polymerization. Typically, the polyester molecules at this stage have 20–30 links and molecular weights of the order of 2,000.

e. Styrene Ratio. The addition of styrene to the reacted polyester renders it a liquid at room temperature and provides the means for its final cure via the styrene cross-linking reaction mechanism. The viscosity of the uncured resin is a function of the degree of polymerization and the styrene content. A resin of a higher degree of polymerization will require more styrene to reduce the viscosity. Viscosity is, of course, critical in determining the wet-out time of the resin in contact and resin injection molding. Higher styrene contents lead to higher exotherms, greater shrinkages, and reduced tensile strengths. A working compromise is found to be about 30% styrene in most formulations.

Styrene has a low boiling point and high vapor pressure, and in open mold operations it is inevitable that some evaporates into the atmosphere. The cure exotherm assists this process, and since styrene vapor constitutes both a health and fire hazard, adequate ventilation must be provided. Since regulations governing the permitted styrene vapor levels have recently been tightened, there is a trend toward the use of less styrene in contact molding formulation or toward its substitution by alternative monomers, e.g., diallylphthalate (DAP).

f. Additives and Cure Promoters. Promoters are usually organic peroxides—benzoyl peroxide (BPO), the ketone peroxides (especially methyl-ethyl-ketone peroxide—MEKP), and cyclohexanone peroxide are typical. When heated, these chemicals produce free radicals which initiate the styrene polymerization and the linking to unsaturated sites in the resin. They do not function at ambient temperatures unless an additional accelerator or promoter is added. Several accelerators are available, but cobalt napthanate and dimethylamine (DMA) are most commonly used. The "activity" of the initiator-accelerator system determines the rate of cure and, thus, both the gel time and peak exotherm. Their action can be further modified by the use of retarders such as α-methyl styrene and inhibitors such as hydroquinone or p-tert-butyl catechol (PTBC). These additives are generally used in very small proportions, typically less than 1% of the resin weight. Since the peroxides are very unstable, they constitute a fire and explosion hazard if mishandled. In particular, they react explosively if mixed directly with the accelerators. For some applications it is convenient to divide the resin into two equal batches, adding the initiator to one and the accelerator to the other. These two batches are then mixed 1:1 immediately before molding. This premixing gives long pot lives to the bulk mixtures and eliminates the need for tricky mixing operations on the shop floor. Another system is to pump the two batches to the molding station, where they are mixed and dispensed in a special metering gun. This system can be used most effectively in the spray-up and hand lay-up processes.

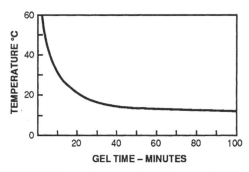

FIGURE 3.5-3. Temperature vs. gel time for a polyester resin.

The basic aim of the final compounding operation is to control the rate of cure and the exotherm to give an adequate pot life and the desired rate of cure without excessive heat build-up. Thus, a much more active mix can be used for making small thin moldings in metal molds than for thick sections in GRP molds. Excessive exotherm energy can result in "styrene boil," with foaming of the resin and charring and degradation in extreme cases.

Resin Cure

The majority of resins supplied for contact molding operations are designed to cure at room temperature, and careful control must be exercised if optimum results are to be achieved. In the first place, the rate of cure is very sensitive to small variations in temperature, as shown in Figure 3.5-3. It is recommended that curing never be undertaken at temperatures below 15°C, and ideally, the shop should be maintained at constant temperature and controlled humidity. Cure in the open shop is achieved at temperatures of 20–25°C. When special tunnels or bags are utilized, a higher temperature can be used and the cure time reduced.

The resin must be formulated with initiator and accelerator to give adequate pot life (Figures 3.5-4a and 3.5-4b) and a minimum gel and hardening time consistent with keeping the exotherm within bounds. The heat generated by the exotherm also speeds up the cure. For maximum pot life, it is beneficial to keep the bulk resin at temperatures below 15°C.

The three significant points during the cure cycle are the gel time, the period beyond which the resin is no longer liquid and is unworkable; the hardening time, the interval before the molding and mold can be sepa-

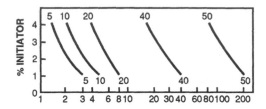

HOURS TO REACH BARCOL HARDNESS AT 20°C

FIGURE 3.5-4a. Hours to reach Barcol hardness at 20°C.

rated; and the maturation time, the point when the laminate has developed its full properties. The hardening time is typically 5–10 times the gel time and may range from about 30 min to several days. Maturation time is usually several weeks at 20°C.

Although the molding can be handled during the period after hardening, it should not be exposed to harsh chemical environments before the full maturation period has elapsed. Maturation can be speeded up by postcuring at temperatures of 50–80°C (Figure 3.5-5). Likewise, the earlier stages of the cure can be accelerated by heat assistance. This heat must be applied with caution because there is a danger of a runaway exotherm. As previously shown in Figure 3.3-6 (section 3.3), the peak exotherm occurs shortly after gelation. It is inadvisable to apply any additional heat before this point. Thereafter, the cure can be speeded up considerably by gradually raising the temperature to 50 or 60°C. After the molded parts have been trimmed and removed from the molds, a postcure at temperatures up to 80°C may be used to accelerate the maturation process. This postcure generally induces higher heat distortion and chemical resistance and helps disperse any free styrene remaining in the molding.

Heat-assisted cure is most conveniently applied in automated production line operations where the mold can circulate through temperature-controlled tunnels during cure. Heating can be achieved by use of warm air or infrared radiant heaters. In some cases, heating foils are actually buried in the laminate so that it can be heated directly by means of electric resistance.

Vinyl Ester Resins

These resins are similar in general behaviors to the unsaturated polyesters in that they are cross-linked by a free radical-initiated addition polymerization reaction (see section 3.3). The base resin, however, is significantly different, in that it consists of a vinyl-terminated bisphenol A derivative. The cross-linking is achieved via the double bonds in these vinyl end groups. The cured resin, thus, has a structure based on the aromatic bisphenol backbone and, in this respect, is similar in nature to that of some bisphenol-derived epoxy resins or to that of the bisphenol A based polyesters. Thus, the vinyl esters exhibit excellent resistance to water and chemical attack, and they can be formulated to develop a high heat distortion temperature. At present the vinyl ester resins are intermediate in price between polyesters and epoxies; therefore, they would be selected over the polyesters only when superior properties were essential.

3.5.5 Hand Lay-up

The basic principles of the open mold contact molding process are illustrated in Figure 3.4-1 (section 3.4). Variations on this basic process will depend on the type of mold, the size of the molded part, the turn-around time for each molding, and the overall production rate. These factors, together with the performance specification for the molding, will determine the grade of resin chosen, the type of reinforcement, and the level of automatic handling of materials and molds that can be sustained.

For all except the largest moldings, it is desirable that the molds be carried on trucks or conveyors. They can then be moved into a well-lighted and ventilated area for laminating and later moved into a holding area while the cure proceeds and until the parts are ready to be removed from the mold.

For low-rate production of medium-sized moldings, it is often convenient to batch-mix the resin in polyethylene buckets or similar containers. The resin is often supplied with accelerator already added (pre-accelerated); therefore, the final compounding requires only that the initiator be added and fully dispersed. However, this pre-acceleration can be quite difficult, especially with the more viscous resins, since only 1–3% of initiator is commonly used. An alternative technique is to add accelerator to one-half of the resin and initiator to the other in such proportions that a 1:1 mix produces the correct final formulation. This operation is much easier to achieve on the shop floor and is less subject to error. When resin is batch-mixed, it is important to relate the batch quantity to the lay-up rate and the gel time, bearing in mind that the gel time of a bucket of resin is much shorter than that for the same resin spread over a thin laminate, from which the heat is more readily dissipated. Typically, one operator can handle 4 liters (1 gal) of resin in about 20–30 min. A resin formulated to give a gel time of 30 min in a 4-liter batch would then have a gel time of perhaps 2 hr when laminated and require 6 hr before the molding could be separated from the mold. If fast cure is required, it follows that the batch size must be reduced or some form of automatic resin blender/dispenser must be installed. In this case, the resin can be supplied at the lamination point and often sprayed directly onto the mold so that the only handwork necessary is that of positioning the reinforcement and then consolidating and wetting it out by rolling down. The resin/glass ratio may be controlled by metering the quantity of resin supplied to each mold. Commonly, however, it is a question of the operator's judging when sufficient resin has been applied to ensure full wetting out—but without excess resin.

A typical lay-up schedule for a 4 m boat hull might be as follows:

1. Wax the mold.
2. Apply mold release agent.
3. Apply pigmented gel coat resin (usually a thixotropic isophthalic resin) by brush, roller, or spray to a thickness of 0.5 mm (0.02 in). Allow to gel (10–20 min).
4. Apply one layer fine surfacing mat and laminating resin (typically a low viscosity isophthalic—possibly pigmented).
5. Apply one layer 450 g/m² (1½ oz/ft²) CSM and resin.
6. Apply one layer 500 g/m² woven roving and resin.
7. Apply one layer 450 g/m² CSM and resin.
8. Apply one layer 500 g/m² woven roving and resin.
9. Apply one layer 450 g/m² CSM and resin.

This lay-up would give a total reinforcement of 2,350 g/m², which at a glass content of 40% would give a molding thickness of approximately 4 mm (0.16 in). Stiffening sections would normally be incorporated in a molding of this type. These sections can be formed by laying strips of polyurethane foam on top of the skin and laminating further layers of CSM or woven roving over them to form a "hat stiffness" section. A molding

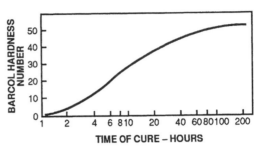

FIGURE 3.5-4b. Barcol hardness versus cure time.

of this complexity might occupy 2 operatives for 4–5 hr for the laminating operation and then require 5–10 hr of cure before being demolded—one molding would require one working day.

Larger parts might take several days to lay up, and a resin with a longer cure time should be chosen so that, although partly cured, the previous layers are sufficiently "green" to allow good adhesion by succeeding layers of resin. With room temperature curing systems, full cure is not normally attained in less than 7 days at 20°C, and significant cure reaction continues well beyond that time.

Maturation should be allowed to proceed to a considerable degree (at least 14 days) before exposing a part to a corrosive environment. Adequate postcure heat treatments (e.g., 10 hr at 70°C) accelerate the maturation process and generally minimize heat distortion and enhance corrosion resistance.

At the other end of the scale, small moldings that can be laid up in a few minutes can be handled on a con-

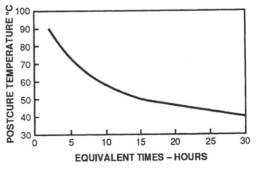

FIGURE 3.5-5. Postcure temperature versus time (e.g., 5 hr at 73°C is equivalent to 15 hr at 50°C). Postcure temperature should not exceed heat distortion temperature.

veyer system. Very short gel times of 5–10 min are used, and the molds are circulated to allow uniform cure (30–60 min). This type of operation lends itself to some degree of automation. Auto-mixing dispensers provide resin freshly mixed with initiator, and often, the resin is sprayed directly onto the mold as the various reinforcements are applied. Where production rates are high, there is considerable incentive to adopt less labor-intensive processes such as spray-up and cold-press (matched-die) molding.

The "Simplex" process utilizes a metered resin supply with a roller applicator so that, as the operator consolidates the molding, liquid resin may be supplied as required. Thus, the resin application and consolidation operations are combined, which is most applicable to medium-sized moldings.

Trimming

In hand lay-up processes the reinforcement is laid over the edge of the mold to give a ragged edge that must be trimmed at some stage. The usual technique is to build a trimming edge into the design of the mold. There is a period during cure—after gelation but before hardening—when this overlap can be most effectively trimmed. For thin laminates, a sharp knife is most effective; for thicker sections, mechanical shears or saws can be used. Trimming at the correct stage of cure saves time and generally produces a neater edge.

Gel Coat

A gel coat is widely used in contact molding. It is formed from a thixotropic pigmented resin specially formulated to produce a high surface finish. It is a self-finish and provides a durable and attractive surface that requires no painting or other finishing operation. In general, the gel resins are less resistant to water and chemical attack than unpigmented and reinforced resin layers. Therefore, there is a move to dispense with the gel coat for the more demanding applications; in small boat manufacture, unpigmented resins are now generally used for below-the-waterline areas. Another consideration is that even the best gel coat will deteriorate over a few years; consequently, modern polyurethane paint finishes are probably a better means of decoration. Other applications in which use of gel coats is inadvisable include those where GRP and metal parts are combined (e.g., in automobiles). Here, the use of col-

ored gel coats would raise very difficult color-matching problems; a better solution is to paint all parts.

3.5.6 Spray-up Processes

Here, the resin and chopped fiber rovings are sprayed simultaneously onto the mold surface. A special spray-head unit receives the liquid resin, adds the correct amount of initiator, and sprays the mix at a carefully controlled rate [2]. At the same time, continuous fiber rovings are fed to the head, chopped into 25–75 mm (1–3 in) lengths and fed into the path of the resin spray (Figure 3.4-2, section 3.4), so that the resin and the fibers are simultaneously directed at the mold in the correct relative quantities. This process clearly improves the degree of control of the process and increases the rate of production in comparison with hand lay-up. However, the process is capable of laying down only a random array of fiber which limits the glass content to about 35%; consequently, properties will be inferior to those obtained from woven roving laminates. After spraying, the laminate has to be consolidated by rolling—as in the case of the hand lay-up.

The main benefits of the process are realized in automated production units. Prepared molds are transported to spraying booths, where they can be manipulated in front of the sprayers to facilitate rapid and uniform deposition of the glass-resin mix. The booths can be properly ventilated to minimize styrene release into the working atmosphere. After spraying, the molds move to rolling-out booths and on to cure, trim, and break-out shops. At all stages, adequate ventilation must be provided.

More sophisticated operations include local reinforcement by hand-laid, pre-tailored woven rovings, which are sometimes pre-impregnated with liquid resin, and stiffened with "hat section" stiffeners, foam, or end-grain balsawood sandwich inserts.

The process is not as versatile as the hand lay-up, and the loss in stiffness and strength realized by the use of random, rather than woven, reinforcement is sometimes larger because of the greater degree of automation that can be exercised.

Production of some simple shapes such as pipes and cylindrical tanks can be completely automated. The spray is directed at rotating mandrels, and all consolidation is mechanical. It is even possible to "program" the spray-head so that thicker layers are put down in critical areas (e.g., around cut-outs).

The very stringent styrene emission requirements have made the spray-up process more attractive, since it is possible to restrict the area where liquid resin must be handled. Since no operator is needed during the cure stage, the molds can circulate through temperature-controlled ventilated tunnels. Environmental control can be concentrated in these areas because human exposure to styrene vapors is restricted to a minimum. All the basic resins can be supplied in grades suitable for the spray-up operation. Lower viscosity grades are preferred, and fast cure rates can be used until a restriction is imposed by the thickness of the laminate. If required, gel coats are applied by spray, brush, or roller at an application station upstream of the main spray booths.

The spray-up process, when applied on a properly automated scale, is the most economical method of GRP fabrication.

3.5.7 References

1. MORENA, J. J. *Advanced Composite Mold Making*. Van Nostrand Reinhold Co., New York, NY (1988).
2. ENGLISH, L. K. "Fabricating the Future with Composite Materials, Part I: The Basics," *Materials Engineering*, 4(9):15 (1987).
3. CHOU, T. W. and J. M. Yang. "Structure-Performance Maps of Polymeric, Metal and Ceramic Matrix Composites," *Metallurgical Transactions*, 17A:1547 (1986).

SECTION 3.6

Compression and Transfer Molding of Reinforced Thermoset Resins

3.6 COMPRESSION AND TRANSFER MOLDING OF REINFORCED THERMOSET RESINS

M. G. BADER

3.6.1 Classical Thermoset Resins and Compounds

The classical thermosetting resins are supplied as solid powders which may be melted by the addition of heat and then chemically cured by a continued supply of heat. In this respect they differ from the addition-cure thermoset resins described in section 3.3 (e.g., the unsaturated polyester and epoxide resins) that are liquid or SMC and BMC materials that are pastes or dough-like.

The classical thermosets comprise a large range of resins: phenolics, amino compounds, alkyds, and glyptal based materials. These materials offer a wide range of property options, but essentially all cure to form intractable highly cross-linked molecular networks. They are harder and more brittle than most alternative systems but offer excellent elevated temperature and electrical insulating properties.

We shall discuss the phenolics as a typical system; they are more properly described as phenol-formaldehyde resins (see section 3.3). They are formed by the condensation of phenol and formaldehyde as shown schematically in Figure 3.6-1. The initial reaction may be to form either a "novolac" (excess phenol and acidic conditions) or a "resole" (excess formaldehyde and alkaline conditions). This initial reaction results in the elimination of water as a condensation by-product. This pre-condensation is a slow process that is carried out in a reaction kettle where the water is evaporated. The resulting novolac or resole, on cooling to room temperature according to the degree of polymerization, may be either a viscous liquid or solid. Typically, molecular weights would be of the order 200–1,000, corresponding to the linking of 2–10 phenol groups. At this stage the resin is fusible (A-stage). Resoles may be further cured by heating to form a cross-linked macromolecule, but novolacs require the addition of a curing agent before further cure is possible.

If required, standard practice would be to prepare the A-staged novolac, or resole, blend, and add a curing agent, usually hexamethylenetetramine (HEXA) together with fillers, pigments, reinforcements, etc. This mix is then hot blended and the curing reaction advanced to the B-stage—the point at which the resin is virtually insoluble in organic solvents but is still fusible when heated. The degree of advancement to the B-stage, which is critical in determining the processing characteristics of the resin, will be further discussed later. The B-staged mixture is cooled and then ground or pulverized to produce the granular solid that is the basis of most molding compounds. The final cure to the C-stage is effected by the application of heat and pressure to the B-staged granules. They first soften and flow under pressure to fill the mold cavity and then cure to a cross-linked intractable solid.

It should be noted that all stages of cure are condensation processes and that water vapor and sometimes other volatiles are evolved as condensation by-products. In processing phenolics, due allowance must be made for the elimination of these volatiles; otherwise, porous or unsound moldings would result.

Control of volatiles evolution may be effected in a number of ways:

a. Advancing the B-Stage—The degree of cure during the B-stage may be advanced to a greater extent. Greater advancement means that the reactions during the final C-stage are minimized; thus, fewer volatiles are evolved. Unfortunately, as the extent of advancement of the resin is increased, its melt viscosity is increased and it becomes more difficult to mold. For the more intricate cavities and where greater extent of flow is required during molding, a less advanced resin will be desirable. Thus, the degree to which volatiles evolution may be controlled by B-stage advancement is limited.

b. Fillers—Fillers are invariably incorporated into

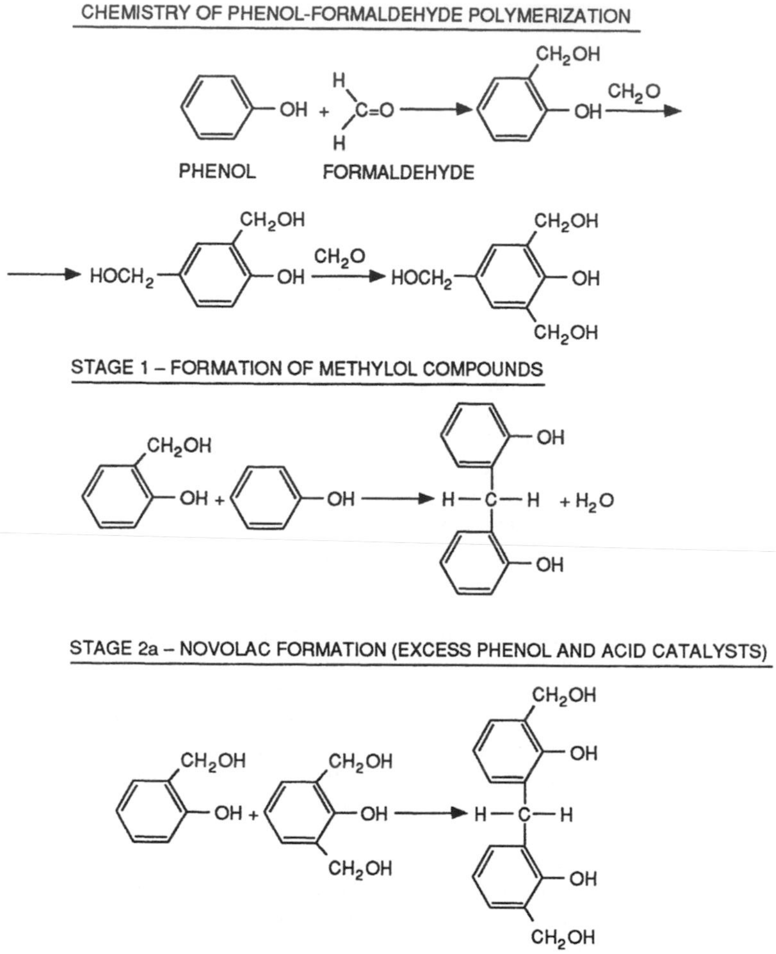

CHEMISTRY OF PHENOL-FORMALDEHYDE POLYMERIZATION

STAGE 1 – FORMATION OF METHYLOL COMPOUNDS

STAGE 2a – NOVOLAC FORMATION (EXCESS PHENOL AND ACID CATALYSTS)

FIGURE 3.6-1. Chemistry of phenol-formaldehyde polymerization.

thermosetting molding compounds. They may be inert organic or inorganic particulate material or reinforcing fillers such as glass fibers. Typically, the total filler content amounts to 40–70% by volume of the compound; thus, fillers serve to control the volatiles evolution by simply reducing the effective resin fraction in the compound. Many compounds include fillers which may absorb water vapor (e.g., wood flour and other cellulose-based materials). The presence of a large volume fraction of filler also provides an escape path for volatiles, at least during the initial stage of cure. Needless to say, fillers have a profound effect on the mechanical and physical properties of the cured resin—a subject which will be further discussed later.

c. Plasticizers and Lubricants—Plasticizers and lubricants may be regarded as processing aids that permit a more highly advanced resin to be used than would otherwise be the case.

Summary

Phenolic resin compounds are based on condensation polymerizing systems. They are invariably highly filled with either inert or reinforcing fillers. Their processing is dominated by the requirement to dispose of the volatiles that are formed during condensation reactions.

3.6.2 Thermosetting Resin Systems

Phenolics

This most important system has been discussed in some detail in section 3.3 and subsection 3.6.1. A large proportion of the phenolic resins produced are used for molding applications, but they are also widely utilized as laminating resins for bonding paper and cloth into board-type products, as adhesives in paints and varnishes, and in several other specialist applications [1].

As the basis for molding resins, they offer excellent properties. They may be manufactured in a wide range of "flow grades" suitable for compression, transfer, and injection molding (section 3.9). A disadvantage is that they are dark colored, ranging from an amber to dark brown according to formulation, which limits their usefulness in certain decorative applications. They are, however, widely used for electrical, radio/television electronics, automotive, and appliance parts, where their combination of mechanical and dielectric properties combined with ease of processing and low cost makes them the first choice.

Amino Resins

These are the resins formed by the reactions of formaldehyde with urea, melamine, or their derivatives (Figure 3.6-2) [1]. In many ways they are similar to the phenolics; they are used in molding compounds, as adhesives, and for the manufacture of the decorative

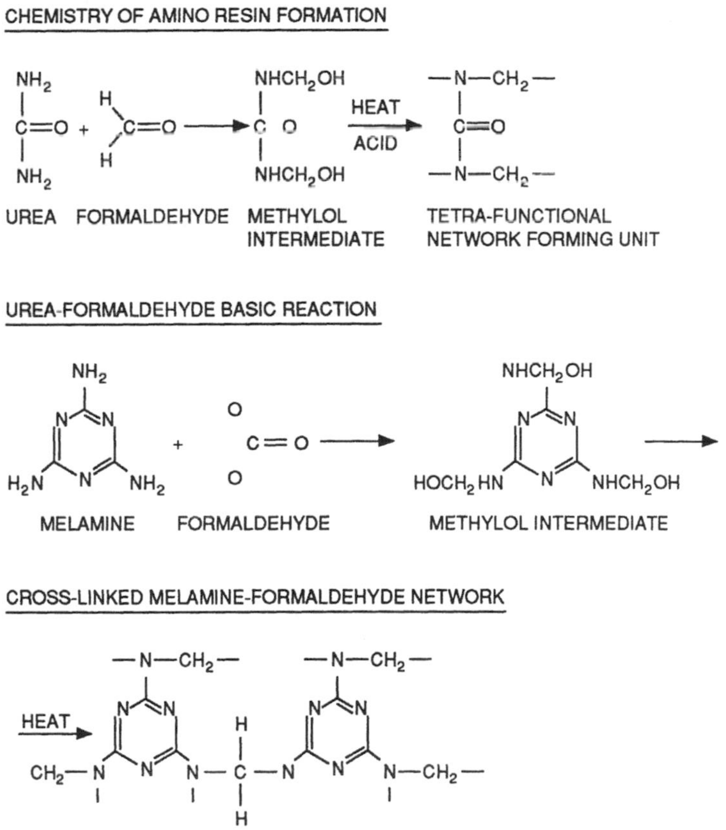

CHEMISTRY OF AMINO RESIN FORMATION

UREA-FORMALDEHYDE BASIC REACTION

CROSS-LINKED MELAMINE-FORMALDEHYDE NETWORK

FIGURE 3.6-2. Chemistry of amino resin formation.

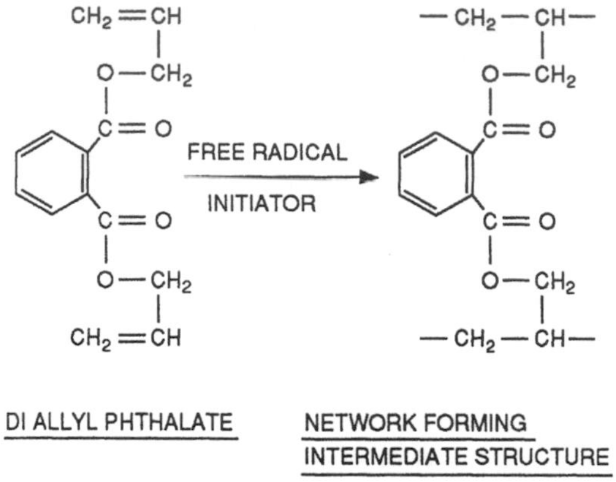

DI ALLYL PHTHALATE

NETWORK FORMING
INTERMEDIATE STRUCTURE

FIGURE 3.6-3. Network forming of di-allyl-phthalate.

paper/cloth/resin laminates utilized in the furniture and appliance industries. As molding compounds, their advantage over phenolics is that they are virtually colorless; therefore, there is no restriction in their use for decorative moldings. The color is, of course, dependent on the choice of fillers. The thermal and electrical properties of amino resins are inferior to those of the phenolics but are generally superior to most thermoplastics.

Alkyd Resins

Alkyd resins are essentially polyester-based resins; the term *alkyd* is derived from the "alcohol and acid" precursors of this family of polymers. They are sometimes termed "glyptal" resins. The unsaturated polyester resins, which are precondensed polyesters dissolved in monomeric styrene (or equivalent monomer), are liquids at ambient temperature and are cured by initiator stimulated addition polymerization cross-linking reactions. The alkyd resins, however, are solids at room temperature but include constituents which cross-link under thermal action without the production of condensates. The most common such ingredient is di-allyl-phthalate (Figure 3.6-3), which is capable of both chain and network formation. Resins based predominantly on di-allyl-phthalate are often termed *DAP resins*. As a group, the alkyd and DAP resins are easy to process; have low shrinkage, excellent electrical resistance, dielectric properties, and moderate elevated temperature properties; and, being naturally almost colorless, can be colored as required.

Epoxides

Although principally used in lamination and casting systems, a range of epoxide-based thermosetting resins is available. Epoxides are generally more expensive than the resins described in the previous paragraphs, but they offer excellent performance under certain hostile environments; good electrical properties, especially surface resistivity and tracking resistance; and very high melt fluidity, which makes them very suitable for intricate moldings and long flow paths.

Special Resins

In addition to the "commodity" resins described under the segments about phenolics, amino resins, and alkyd resins, there are many other possibilities for thermoset systems. Among the more significant ones are a number which could be described as high-performance phenolics that offer generally better high temperature or chemical performance than the conventional phenolics. Examples of this class of resin are the phenol-aralkyl compounds in which phenol is reacted with a range of para-xylene compounds by a Freidal-Crafts reaction to form linear, cross-linkable, aromatic chain molecules. These molecules may be cured with epoxide compounds or hexamine to give cured resins capable of service over 250°C (480°F).

Another group of resins with excellent elevated temperature properties is the polyimide-based materials (section 3.3). Some of these materials can withstand service temperatures of over 350°C (650°F). In general, they are expensive and difficult to mold, but easy-to-process grades with slightly down-graded service properties are now on the market. These materials should be considered whenever service temperatures in excess of 250°C (480°F) are specified.

3.6.3 Fillers and Reinforcements for Thermoset Systems

Fillers are essential components of many thermosetting resin compounds. As mentioned in subsection 3.6.1, they fulfill a number of functions. An essential function is to reduce the amount of resin in the compound. This reduction serves to control the quantity of volatiles produced as the resin cures and, in general, also reduces the overall cost of the compound since the fillers are usually less expensive than the resins. The filler also fulfills a role in controlling the molding shrinkage and the dimensional stability of the molding, and the choice of filler has a considerable influence on the mechanical properties of the molding, in particular hardness and toughness. Electrical and chemical properties may also be affected by the filler chosen (see Volume 2). Finally, the type of filler has a considerable influence on the surface finish and general appearance of the molding. In selecting fillers, due regard must be given to the above considerations. Ideally, the maximum quantity of filler will be used, consistent with the requirements of the molder and the properties required in the molding. There is considerable scope for variation when compounding; for example, one may use more filler with a less advanced resin, or vice versa, to achieve a given flow characteristic on molding. However, the two compounds may show quite different properties in the fully cured condition. Filler contents are typically held at around 50–60% by volume, which is found to give good molding characteristics and satisfactory mechanical and physical properties in the molding. Fine tuning of the flow characteristics of the compound is accomplished by control of the degree of advancement and by use of processing aids such as soaps or waxes which act as lubricants.

There are three principal categories of fillers that are widely utilized: organic, particulate mineral and glass, and fibrous reinforcing fillers.

The organic fillers are very widely used in general purpose molding compounds. They are generally cellulosic in nature and include such materials as wood flour, cotton flock, cotton seed hulls, ground walnut or coconut shells, and cork dust. These fillers are low density, fibrous, and absorbent to some extent. The particle size is typically graded at 80–100 mesh, which means that particles and fibers of up to about 0.1 mm (0.004 in) in size will be present. These particles blend very well into phenolic and amino resins to give low density moldings with good surface finish. The ability of the particles to absorb some water vapor helps to alleviate the release of the water produced during cure. The impact resistance of moldings filled with fibrous organic fillers of this type is generally quite good, but chemical, dielectric, and elevated temperature properties are relatively poor. In particular, they tend to absorb water when they are immersed or placed in a high humidity environment; they ultimately swell and surface degrade. Nevertheless, large quantities of general purpose moldings are produced using these fillers.

Other organic fillers, although quite different in nature, are carbon black and graphite flake. The carbon blacks are sometimes used in quite small proportions; they improve the elevated temperature properties, confer good resistance in some hostile environments, and color the molding black. Graphite flake is used to confer a measure of self-lubrication to the molding for light duty bearing and rubbing surfaces.

The inorganic fillers comprise a vast range of clays, ground rocks of various types, precipitated chalk, silica, mica, asbestos and glass in the form of fibers, flake, and ballotini (small spherical particles).

Precipitated chalk (whiting), talc, and clays are widely used to produce general purpose moldings with good surface finish and moderate electrical and elevated temperature properties. Silica, mica, and glass (flake, ballotini, or fiber) give better electrical and elevated temperature properties and chemical resistance. Mica is outstanding for conferring excellent high temperature electric insulation and dielectric properties. In general, the mineral filled materials are harder and more brittle, and have better dimensional stability than those filled with wood flour and similar organic fillers. They are also more dense and require more careful control of the molding operation if blistering and porosity are to be avoided.

None of the particulate fillers (organic or inorganic) increases the tensile strength of the molding. Commonly, the strength is reduced. Strength improvements can be accomplished, however, by the use of reinforcing fibers that enhance both stiffness and strength and also sometimes improve impact resistance. Formerly, asbestos filler was widely utilized in thermosetting compounds. It conferred modest enhancement of strength with excellent elevated temperature properties. Currently, the use of asbestos is discouraged because of the risk to health from asbestos fiber ingestion; therefore, glass or graphite fiber must be used if a reinforcing filler is required. Glass fiber in the form of very short chopped rovings [1 mm (0.04″)] or milled fibers of even shorter length may be blended into the resin at the final compounding stage. Up to about 40% by volume of glass may be incorporated in this way, depending on whether other particulate fillers are also added. The general trend is to use a minimum of other fillers if glass fiber is used, but individual molding requirements may indicate some other additions for color or surface finish consideration. Excellent mechanical properties, dimensional stability, and resistance to elevated temperatures

Table 3.6-1. Properties of typical thermoset molding materials.

Base Polymer	Filler	Filler Content % (wt) (approx)	Density g/cm³	Molding Shrinkage %	Cure Temperature °C (°F)	Flexural Strength MPa (ksi)	Tensile Strength MPa (ksi)	Impact Energy ft.lb/in	Water Absorption % (24 hrs)	Heat Distortion Temperature °C (°F)
Phenolic	Wood Flour	30	1.34	1.0	165 (330)	83 (12)	55 (8)	0.4	3	140 (280)
Phenolic	Cotton	30	1.38	0.7	165 (330)	90 (13)	70 (10)	2.4	5	135 (275)
Phenolic	Mica	60	1.90	0.2	170 (340)	83 (12)	35 (5)	0.4	0.5	170 (340)
Phenolic	Asbestos	60	1.84	0.4	165 (330)	75 (11)	40 (6)	0.7	2	145 (290)
Phenolic[2]	Glass Fiber	60	1.8	0.2	165 (330)	100 (15)	70 (10)	0.5	0.2	190 (375)
Amino	Wood Flour	30	1.5	1.0	165 (330)	90 (13)	60 (9)	1.4	2	120 (250)
Amino	Cellulose	30	1.5	1.0	165 (330)	95 (13.5)	55 (8)	1.5	1.5	100 (210)
Melamine	Cellulose	30	1.5	0.8	165 (330)	100 (15)	60 (9)	1.5	1.5	140 (280)
DAP[1]	Glass Fiber	50	1.85	0.1	170 (340)	125 (18)	65 (9.5)	15	0.4	200 (380)
Polyimide[1]	Glass Fiber	60	1.95	0.05	220–250 (425–480)	248 (36)	145 (21)	22	0.2	315 (600)
Polyimide[2]	Glass Fiber	50	1.85	0.4	220–250 (425–480)	70 (10)	55 (8)	0.4	0.1	315 (600)

[1]Compression molding grade (long glass fibers).
[2]Injection molding grade.

and to hostile chemicals may be obtained from glass-fiber compounds, but, in the case of non-spherical fillers (e.g., fibers and platelets or flakes), flow during molding will induce an alignment of the fillers with resultant anisotropy in the molding. This effect is most marked in thermoset compounds when glass fiber, glass flake, and mica fillers are used. The properties of typical thermoset materials are given in Table 3.6-1.

3.6.4 Compression Molding

The basic principles of compression molding have been set out in subsection 3.4.1 and Figure 3.4-5. To summarize, a pre-weighed charge of solid thermoset (loose or pelletized granules) is placed in the hot mold. The two halves of the mold are brought together and a controlled pressure applied. The compound first softens and then flows under the influence of the pressure to fill the mold cavity. Heat and pressure are maintained; when the material has cured, the mold is opened and the hot molding is ejected.

We will now discuss compression molding of thermosets in more detail. We need to consider

a. Materials selection and pre-molding operations
b. Molding
c. Mold and cavity design
d. Molding machinery
e. Post-molding operations

At each stage we will consider the influence of the available options on the properties of the molded part and on the production process.

We first generalize the objective: to produce parts of adequate and consistent properties at the highest possible rate and in the most cost-effective manner. As always, some of these objectives tend to be mutually exclusive so that the final materials selection and processing specification reflect the practical compromises which must be accepted.

Pre-Molding Operations and Materials Selection

It is assumed that the designer will have specified the basic category of materials to be molded (e.g., a glass fiber filled phenolic). The specification can then be further refined by consideration of the size, complexity, and property specification of the molding. As a general rule, the properties obtained are better if advanced resins are being used. However, they may not flow adequately to fill a complex cavity. To some extent, this inadequacy might be countered by increasing the molding pressure. But one must be cautious lest the latitude available during the molding process (e.g., temperature, time, pressure) be reduced beyond the capabilities of process management (or process/materials variability). If this happens, erratic molding quality and high reject rates will inevitably result.

Having established the material for the molded part, the next stage is to ensure that a consistent charge is delivered to the press; the charge must be metered or weighted. A common practice is to pelletize the loose granular material in an automatic pelletizing press. The pellets are typically cylindrical, 10–60 mm (½–2½ in) diameter and 5–25 mm (¼–1 in) long. Ideally, the pellet weight should be controlled so that 1–5 pellets (typically) form a charge. This control reduces the possibility of charge weight variations due to human error and allows for automatic charging.

The charge must be dry. The molding compound is normally supplied dry in sealed containers, but it may be very hygroscopic and pick up unacceptable quantities of water during normal handling. The best practice is to dry the charge immediately prior to molding. This drying may be accomplished by use of a circulating air drying oven (or tunnel), a vacuum drying oven, or R.F. (radio frequency) heating.

The drying operation is often combined with a preheating operation whereby the pellets are heated to 50–70°C (120–160°F) before charging. This preheat not only dries them but also reduces the necessary heat-up time in the mold which speeds up the molding cycle.

Automatic charging equipment is available for use in high production rate, fully automated, molding machinery. This equipment accepts the molding compound in bulk form (powder, granules, and sometimes paste), dries it, preheats it, if required, and delivers metered doses to each mold cavity. This type of equipment would be justified only for long production runs; it is most often applied to large multicavity molds making relatively small components.

Molding

The molding operation requires that the charge be softened, compressed, and cured to give a good quality part in the shortest possible time.

It must be understood that the charge, whether pre-pelletized or loose fill, has a low thermal conductivity. Most compounds will soften at temperatures of 80–110°C (180–230°F) but will cure rapidly at temperatures above about 140°C (285°F). Optimum cure temperature for phenolics is usually in the region of 170–190°C (340–375°F) and the mold will be heated to maintain this temperature. Thus, the charge is rapidly heated through the softening range and into the curing range where it is in contact with the mold surface, but parts of the charge remote from a hot surface may take from several seconds to minutes to heat up into the softening temperature range. If too much pressure is applied before most of the charge has softened, there is a danger that only part of the charge will flow—perhaps into narrow sections of the mold where it will be heated rapidly to the cure temperature and, having cured, constitute a blockage to the flow of further material. This blockage effectively prevents the most remote regions of the mold from being properly filled. The remedy is to close the mold with minimal pressure after charging and allow a short warm-up period before the pressure is applied at a controlled rate up to the full molding pressure required. This remedy ensures uniform flow and good mold filling. It must be emphasized that these precautions are necessary only in the case of the more difficult moldings involving long flow paths and highly advanced (low flow) resins. In many instances, it is possible to apply full pressure just as soon as the mold has been charged.

The principal processing variables during compression molding are mold temperature, molding pressure, and ram speed (or rate of application of pressure). Higher temperatures encourage faster softening and faster cure and, within limits, will induce a higher degree of cure in the resin with a higher glass transition and heat distortion temperature. However, a higher rate of cure is accompanied by a higher rate of volatiles evolution, and there is a danger that these volatiles might not be able to escape before being entrapped in the cured resin. This entrapment leads to voids and blisters in the molding. There is also a danger of premature cure before the mold is completely filled, which leads to short moldings and also to weak weld or knit lines in complex parts. If the molding is ejected at too high a temperature, there is a greater possibility of distortion during the cool down period. For these reasons, it is often advantageous to operate the mold at temperatures below the maximum recommended for the resin. The penalty of a longer molding cycle is compensated for by a better quality part.

The molding pressure must be adequate to ensure complete filling of the mold, but too much pressure can

(a)   **CORE PIN REMOTE FROM GATE –
SEVERE WELD LINE**

(b)   **CORE PIN CLOSE TO GATE –
LESS SEVERE WELD LINE**

(c)   **DOUBLE GATING – SEVERE WELD LINE
WHERE FLOW STREAMS IMPINGE**

(d)   **DOUBLE GATING – WELD LINE FORMS EVEN**

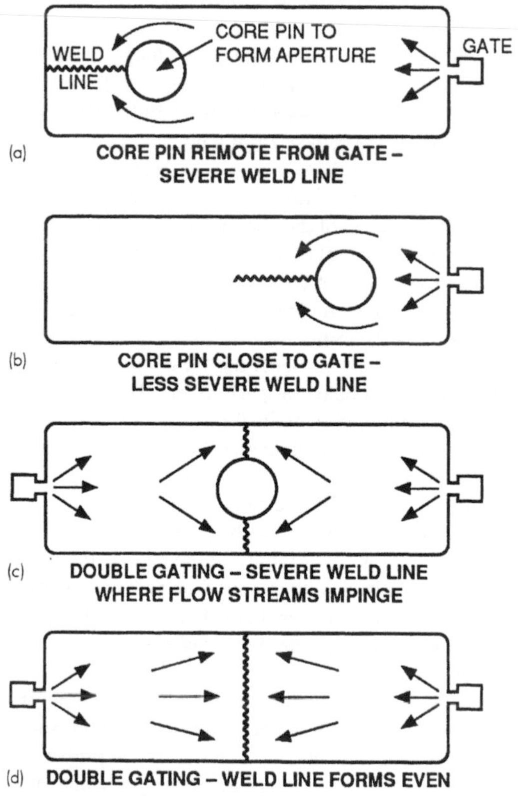

FIGURE 3.6-4. Weld line formation in transfer and injection molded components.

also be detrimental. Actual molding pressures vary quite widely; they are typically within the range of 5–20 MPa (50–200 bar or 700–3,000 psi) for straightforward compression moldings. More intricate moldings require higher pressures, as do the more intractable molding compounds. There is, therefore, a tendency to use easy flow (low viscosity) compounds wherever possible to reduce the stiffness requirements of the mold and allow smaller capacity presses to be utilized. Mold cavities may distort, and there is a tendency to obstruct venting paths so that pockets of volatiles are entrapped at high pressure. These pockets subsequently burst as blisters or form internal cracks as the pressure is relaxed when the mold is opened. Molding pressures will need to be higher when low flow (high viscosity) resins are molded, but these resins usually give rise to less volatiles.

The rate of compaction of the charge is also important. At high ram rates, quite apart from the question of warm-up time discussed previously, irregular flow patterns can be induced that result in unpredicted and variable knit lines in the molding.

For trouble-free molding, temperatures and pressures should be selected that fall well within the overall acceptability envelope for the operation. Care must be taken to ensure that molds are uniformly heated and that temperature control is adequate (typically $\pm 2°C$ or $\pm 4°F$ would be considered adequate in most cases). Pressure and ram rate control usually present fewer problems than temperature control.

Mold and Cavity Design

The facts that molding temperatures are generally above 160°C (320°F) and pressures above 5 MPa (50 bar or 700 psi) demand that molds be manufactured from steel. However, in some non-demanding circumstances, zinc-based alloy has been used successfully.

The basic equipment for compression molding consists of a pair of matched half molds that close to form the cavity. The molds must be heated either by contact with heated platens on the molding press or by heaters submerged in the mold itself. Suitable heating media are steam, circulating oil, and electrical resistance heaters. As a general rule, direct heating of the mold results in a more even temperature distribution than heating by conduction via the platens, but the mold is more complex and expensive. Also, mold changing becomes a more complicated routine. Heating by a circulating fluid is particularly attractive for large installations where many molds may be heated from a single recirculating fluid (oil) system. Electrical resistance heaters tend to become more troublesome over long campaigns.

The mold cavities must be finished to dimensions which allow for cure shrinkage during molding or post-molding operations. This shrinkage can be very small (less than 0.5% linear) with many mineral filled compounds but is often not uniform in all directions. Therefore, it is prudent, when very accurate control of dimensions is required, to initially form the cavity slightly undersized and then to establish the extent of shrinkage before machining to the final dimensions. This final machining will also establish the degree of perfection of the molded surface. Ideally, the cavity should present a hard, highly polished surface. This finish can be achieved by forming the mold in a non-shrinking die steel which is hardened and tempered before the final machining operations. The cavity is often hard-chromium plated to further enhance surface finish and wear resistance.

Molding cavities must be designed with no undercuts and, ideally, a minimum draft angle of 1–2° which ensures easy ejection. It must be emphasized that many compounds, even when fully cured, are relatively soft and flexible at the demolding temperature and may be distorted by excessive ejection forces. From a processing point of view, a part should be designed with uniform wall thickness, which, however, is often not compatible with the geometric requirements of the component, but care should be exercised to avoid sharp changes from thick to thin sections and vice versa. Where such change is unavoidable, the cavity should be charged so that material flows from thick to thin section — not the reverse. Metal and/or ceramic inserts may be successfully incorporated into the thermoset moldings, but care must be taken to minimize the effect of the differential thermal strains that will be induced around inserts.

Undoubtedly, the most important single feature of cavity design is adequate venting of the cavity to allow the volatiles to escape harmlessly during the molding operation. Venting is best achieved by grinding venting grooves into the mating faces of the mold. These grooves might typically be 0.01–0.05 mm (0.0004–0.002 in) deep and up to a few millimeters wide. In many cases, these venting passages may be incorporated within the overall design or decorative embellishment of the molding, e.g., a textured surface. Further vents can be formed by grinding flats on ejector and core pins. The general principle is to vent all extremities of the mold with special attention to the thicker sections where more volatiles will be evolved. It must be noted that the extent of venting required will vary according to the type of resin being molded. If the venting passages are too generous, then there is the possibility of the charge "flashing" into the vents, which in turn may interfere with demolding.

The other very important requirement of cavity design is the avoidance and control of weld/knit lines. Wherever the charge flow is split into streams that subsequently rejoin (e.g., by flowing around a core-pin), there will be a weld line. Inevitably, the two fluid streams tend to combine smoothly so that a surface of weakness is formed. This weakness tends to be more pronounced if the flows are slow and the resin is partly cured when the impingement occurs, as is shown in Figure 3.6-4. It is necessary therefore to study the flow path of the charge in the mold [2] and to try to ensure that, if weld lines cannot be avoided, they occur in non-critical parts of the molding and that molding conditions are set to ensure adequate flow in the weld line area to facilitate the formation of a sound join. For instance, the configuration shown in Figure 3.6-4(b) is better than 3.6-4(a), while in 3.6-4(c) the direction of the weld line has been altered by the changed gate arrangement.

To summarize, the mold cavity and mold must be uniformly heated, and the cavity must be designed with due regard to shrinkage, surface finish requirements, venting, and possible weld line problems.

Molding Machinery

The range of compression molding machinery is vast, ranging from simple hand-operated presses of around one metric ton (2,200 lbf) closing force to giant automated presses of over 1,000 metric tons. The simplest requirement is a platen press with a means for heating either platens, mold halves, or both. Hydraulic operation is usually the most convenient. Since molding pressure requirements are in the order of 10 MPa (100 bar or 1,500 psi), each metric ton of press capacity is equivalent to a projected cavity area of 10^{-3} m^2 (10 cm^2) [or each U.S. ton (2,000 lbf) of press capacity — projected cavity area of 1.3 in^2]. Press capacity can sometimes be increased by using the multiple daylight principle of several sets of platens in series being operated by a single hydraulic ram. However, the more usual practice is to use larger presses with multiple cavity molds (for small components, often tens of cavities per mold). High production rates are secured by minimizing the mold cycle time. Typically, modern high cure rate compounds will cure in 1–10 minutes, depending on wall thickness of the part. To this time must be added the press operating time and the time consumed in charging and demolding. For small components that cure quickly, a rapid press cycle with automatic charging and demolding will pay dividends, but when cure cycle time is over two minutes, there is often little purpose in having fast operating presses. However, automated charge and demolding will still reduce labor costs.

Post-Molding Operations

It is always a tendency to reduce the total molding cycle time to a minimum in order to increase the production rates that might result in inadequate cure. Sometimes it is advantageous to deliberately undercure in the mold and then to postcure in an oven at a higher temperature. Postcure can also be advantageous when very high dimensional stability is required in service, or, in the case of suitable molding compounds, a higher elevated temperature resistance is required. Many compounds, especially phenolics, can be coaxed to give better elevated temperature performance by heating slowly to progressively higher curing temperatures. As an example, a compound which can be demolded after say 5 min at 165°C (330°F) might be postcured at 175°C and then 200°C (350°F and 390°F) for periods of up to 72 hr to induce the highest possible degree of cure as reflected by the heat distortion temperature. Such treatments are much too long to be carried out in the mold, and attempting an initial cure at the higher temperatures would almost certainly result in blistering. Blistering occurs because, during the long soak periods, not only does the cross-link network develop, but also volatiles trapped in the molding are allowed to diffuse out. Without postcure, a part at elevated temperature service, say 150°C (300°F), might show progressive shrinkage over a long period.

Postcure may be carried out in continuous tunnel ovens, in air circulating batch ovens, or in heated oil baths. The latter alternative gives more support to delicate moldings that might otherwise distort under gravity when heated at high temperatures.

The only other post-molding treatment of consequence is mechanical trimming, deflashing, or deburring, which, for small robust components, is carried out by tumbling in a suitable medium, i.e., mild abrasives such as nut hulls or pumice.

3.6.5 Transfer Molding

The basic principle of this process is described in section 3.4 and Figure 3.4-6. The essential difference between compression and transfer molding is that in the latter case the charge is placed in the transfer cavity, rather than the mold cavity, from which it is transferred by the action of the ram. From a practical viewpoint, conditions are relaxed, in that an accurately metered charge is no longer necessary since, in the transfer process, the molding cavity dimensions are fixed when the mold is closed. Conversely, in compression molding, the extent of mold closure is controlled by the charge volume. It is also clear from Figure 3.4-6 that excess charge must be placed in the transfer cavity and that this surplus, together with the runners, is destined to become scrap.

The mold and transfer cavity are heated. In principle the transfer cavity need be heated only to the softening temperature of the charge and the mold to the cure temperature, but it is common practice to heat both to the cure temperature in order to simplify mold design and process operation. The transfer molding cycle is as follows:

1. The mold is closed, the transfer cavity opened, and the charge placed in the cavity.
2. The ram is entered into the transfer cavity—the dwell period is set to allow the charge to heat up.
3. The ram is operated to transfer the charge into the mold cavity through the runner system.
4. The ram pressure is maintained while the resin cures.
5. The ram is withdrawn. Dovetail in the ram causes excess charge and runner (which is also cured) to be withdrawn with the ram. This excess is ejected and scrapped.
6. The mold is opened and the part is ejected.

In comparison with the compression molding process, the molding compound is required to flow more, which confers the benefits of a more homogeneous and more highly plasticized melt. There can be some flow-induced orientation in the molding, especially when fibrous or plate-like reinforcement or filler is used [3].

Molding tolerances are held to tight limits without the need for very accurate charge metering. However, the excess charge in the runners and transfer cavity is waste since the cured thermoset is usually not suitable for regrinding and reuse. Some compounds are available, however, which will accept a small proportion, about 5%, of regrind material. In general, between 2 and 10% of the material charged is lost as waste in this process, whereas none need be lost in simple compression molding. Resin transfer molding is also attractive when complex multi-impression tooling is used, since a single charge placement may be used to fill perhaps twenty individual cavities through a complex runner system.

To summarize, the transfer molding process will in many cases allow more accurate control of dimensions and the achievement of better physical properties than in simple compression molding. Compounds with easy flow characteristics are required, and a proportion of waste is inevitable.

It should be noted that most of the arguments which favor transfer molding over compression molding are even more valid when applied to injection molding of thermoset materials. This technique, which is being applied to thermosets on an increasing scale [4], is discussed in section 3.9.

3.6.6 References

1. BRYDSON, J. A. *Plastics Materials*. 4th ed., Butterworth Scientific, London (1982).

2. EDULJEE, R. F. and J. W. Gillespie, Jr. "Process Induced Fiber Orientation and Weld-Line Studies," Center for Composite Materials, Report CCM-85-12, University of Delaware, Newark, DE (1985).

3. GILLESPIE, J. W., JR., J. A. Vanderschuren and R. B. Pipes. "Process Induced Fiber Orientation: Numerical Simulation with Experimental Verification," *Polymer Composites*, 6(2):82 (1985).

4. ENGLISH, L. K. "Fabricating the Future with Composite Materials, Part I: The Basics," *Materials Engineering*, 4(9):15 (1987).

SECTION 3.7

*SMC and BMC Press
Molding Technology*

3.7 SMC AND BMC PRESS MOLDING TECHNOLOGY

A. B. ISHAM

3.7.1 Introduction

The basic principles of compression molding were outlined in section 3.4 (Figure 3.4-5), and standard compression molding processes and resin systems were more fully treated in section 3.6.

Two successful variants of compression molding in the composites industry involve the use of Sheet Molding Compound (SMC) and Bulk Molding Compound (BMC). They offer the automotive, appliance, and equipment industries the capability for high volume production [1].

SMC is prepared by chopping continuous strand roving onto a plastic film that has previously been coated with a filled polyester resin paste. The paste and glass reinforcements are gently "kneaded" together, and a sheet product is formed (see Figure 3.7-1). Rolls or boxes of SMC are stored until the viscosity has increased to a predetermined level and the material is rubber-like. This step can take from a few hours to several days, depending on the type and quantity of thickener used in the formulation.

BMC is prepared by thoroughly mixing chopped strands with a filled polyester resin paste that can be used in bulk form or extruded into a rope for easier handling. BMC can be molded immediately after mixing or may be stored like SMC in order to increase the viscosity to specified levels.

For molding, both SMC and BMC are preweighed to specific size charges and placed into the mold prior to application of heat and pressure. Pressure can be varied from 3.4 to 13.8 MPa (500 to 2,000 psi). The length of cure depends on temperature, resin, resin initiator, and part thickness. Cure temperatures range from 132 to 160°C (270 to 320°F). At these temperatures, the cure cycle can vary from 1–5 min. In thick moldings, cure cycles must be increased so that the exothermic chemical reaction does not cause material degradation.

3.7.2 Selection of Materials

Resins

Unsaturated polyester or vinyl ester resins are the predominant polymers used in SMC and BMC. These resins are described more fully in section 3.3. The initiators for these resins are chosen so that the mixed system will have reasonably long shelf life at ambient temperature and will cross-link rapidly at the mold temperature. Recent developments of "low shrink" and "low profile" polyester resin systems are being vigorously employed in the industry to meet end use requirements. Low shrink systems contain up to 10% thermoplastic polymers by weight of total resin, while low profile systems contain from 10 to 15% [2].

Low shrink systems are easily pigmented, reduce surface waviness, and achieve a mold shrinkage as low as 0.1%. Low profile systems are not readily pigmentable but offer smoother surfaces for painted applications such as body panels in the automotive industry. With low profile systems, it is possible to achieve 0 to 0.05% mold shrinkage. Mold shrinkage is defined as the difference between the linear dimensions of the cold part and the cold mold. This measurement includes the combined effects of polymerization as well as the difference in thermal expansion between the molded composite and the steel tool.

Most SMC compounds and many BMC compounds contain an alkaline earth oxide or hydroxide to chemically thicken the compound prior to molding. Magnesium oxide, MgO, or magnesium hydroxide, $Mg(OH)_2$, are most commonly used. The thickening of the polyester resin results in a marked increase in viscosity with time but does not affect the ability of the resin to be cured to a cross-linked thermosetting polymer through the action of the initiator, when heated in the mold.

133

SMC-R MACHINE

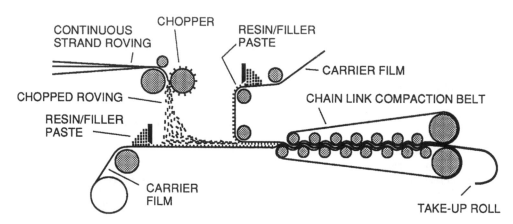

FIGURE 3.7-1. Processing equipment for SMC containing random chopped strand reinforcement.

Fillers

SMC and BMC compounds normally contain fillers to reduce cost and to control flow during molding. For automotive applications, calcium carbonate is used extensively due to its low cost and its ability to provide smooth molded surfaces. A typical automotive compound may contain 150 to 200 parts of calcium carbonate per 100 parts of resin. For many other applications, fillers are chosen to impart characteristics not obtainable with calcium carbonate. For example, alumina trihydrate is used to improve electrical properties and to reduce flammability. Alumina trihydrate releases 34% of its weight in water vapor when it is burned.

Reinforcements

Continuous rovings or chopped strands are the reinforcements of choice for SMC and BMC compounds. These reinforcements have been described in section 3.2.

3.7.3 Compounding Technology

SMC Compounding

Sheet molding compound is produced by an automated, continuous flow process. SMC processing machines can be equipped to produce compounds with random chopped fibers (SMC-R), continuous unidirectional fibers (SMC-C), or combinations designated as SMC-C/R [3]. Using these machines, glass-fiber content can be controlled to within 2% or better. Machine size varies with the desired width of the end product. Most machine manufacturers offer a range of widths from about 0.6 to 1.5 m (2 to 5 ft) with 1.2 m (4 ft) being most common. Another variation, thick molding compound (TMC) can be made in sheets up to 50 mm (2 in) thick, as compared to 6 to 13 mm (¼ to ½ in) for conventional SMC material.

The basic machine configuration is shown schematically in Figure 3.7-1 [4]. Using these machines, the resin paste is compacted between two "carrier films" of polyethylene and/or nylon. The resin paste—containing the thermosetting resin, filler, initiator, mold release, and thickening agents—is uniformly metered onto the lower film. Glass-fiber roving is fed in continuous strands through a chopper assembly that cuts the fiber to a desired length. This length is normally from 13 to 51 mm (½ to 2 in). The chopped fibers drop onto the resin paste-coated film. The top carrier film, coated in a similar manner, is fed onto the moving belt to form a continuous sandwich of glass and resin. This sandwich is compacted under controlled pressure and then taken up on standard packaged-size rolls.

A single SMC machine can provide material requirements for many molding presses. At 100% efficiency and continuous operation, a single machine with a width of 1.2 m (4 ft) can produce 11 million kg (25 million lb) of molding compound per year. Of course, normal machine shutdown periods for formulation changes, cleanup, and preventive maintenance will reduce practically available volumes.

With that background, the details of the process and its control to assure a product with desired characteristics will be considered next.

Resin Paste Mixing

Before resin paste is delivered to the machine, the resin paste ingredients must be properly mixed. There are three basic paste mixing techniques — batch, batch/continuous, and continuous.

Batch Mixing. In batch mixing, all raw materials are mixed in a single mixing unit. The size of the SMC machine usually determines the amount of resin paste needed. For example, a small machine may require batches of only 19 to 38 liters (5 to 10 gallons), whereas a large machine may need 38 to 76 liters (10 to 20 gallons) of resin paste even for short production runs or evaluation of experimental formulations. Where large quantities of material are to be produced, a manufacturer can easily justify the cost of an automated mixing system.

Batch/Continuous Mixing. Batch/continuous mixing normally employs two tanks — one to hold the thickenable resin batch and one to hold a non-thickenable mix that contains the thickener itself. A metering pump, or cylinder, which insures proper proportions from the two tanks, pumps the two mixes simultaneously through a static or dynamic mixer to the metering sections of the SMC machine. Another possible setup is to have thickenable resin paste in one tank and only liquid dispersed thickener in the other. This setup still requires batch mixing equipment, but reproducibility of the resin paste delivered to the machine improves markedly. The primary advantages of batch/continuous mixing over batch mixing are greater reproducibility in resin paste thickening, higher material utilization, and reduced labor costs.

Continuous Mixing. In continuous mixing, predetermined amounts of the liquid ingredients are individually pumped into a continuous mixer. Dry ingredients can be preblended or fed individually by automatic metering equipment. The amount of thickened resin

paste in the continuous mixer can be kept to a minimum for delivery to the SMC machine. Reproducible pumping and metering rates are used, allowing better control of resin mix.

A significant advantage over batch/continuous mixing is the elimination of a separate resin paste mixing facility. It also provides higher material efficiencies than either batch or batch/continuous mixing, since a change of resin formulation or normal cleanup operations waste less material.

Glass-Fiber Reinforcement Section

The part of the machine where chopped fibers are added includes the creel, the guide and tensioning bars, and the chopper assembly. This equipment is designed to deposit a uniform and controllable blanket of chopped strands onto the carrier film coated with resin paste.

From the creel station, continuous strands of glass fiber roving are pulled through the chopper assembly. Tension bars are provided as part of the feed mechanism, enabling the chopper to pull the continuous strands without twisting or tangling. The strands are directed through ceramic guide eyes, which also minimize abrasion of the rovings. When the creel station is located some distance from the machine, the strands can be fed through small diameter stainless steel tubes rather than guide eyes. These tubes will prevent tangling or twisting over the longer feed length.

The chopper assembly includes a cutter roll and an opposing cot arbor for the cutter roll to work against. The cutter roll has a series of razor-sharp blades adjusted to cut specified lengths of glass. The cot arbor is made from natural rubber, urethane, or some other elastomer with a hardness of about 60 Shore A. The chopper assembly is placed from 45 to 75 cm (18 to 30 in) above the machine's conveyor belt. The cutting roll can be adjusted for fiber lengths from 13 to 75 mm (0.5 to 3 in). Fibers longer than about 75 mm (3 in) tend to cluster and to become oriented in the direction the conveyor is moving. Fibers shorter than 13 mm (½ in) may stick between the chopper blades, requiring an air assist to blow them loose.

Paste Metering Section

As seen in Figure 3.7-1 and described earlier, there are two paste metering stations on each SMC line. The mixed paste is fed to these stations, and adjustable doctor blades are set to meter a predetermined thickness of paste onto the carrier film. The doctor blades are positioned vertically to the conveyor belt and are beveled to a fine edge away from the paste flow. This bevel reduces the variation in the amount of resin paste metered on the carrier film due to viscosity and temperature changes. The amount of resin paste which is deposited on the carrier film is affected by both the doctor blade opening above the film and the total amount of resin paste behind the blade. As the level of paste behind the blade decreases, so does the feed rate. Thus, it is advisable to supply the paste continuously from the mixing station in order to maintain a constant level.

Compaction Section

In the compaction section, the film "sandwich" passes between two steel chain link belts, which are driven by a pair of steel rollers. The belts force the resin paste layers into the glass-fiber blanket to achieve uniform wet-out of the product. The series of rollers that the belts pass between is usually attached to air cylinders to permit variable applied pressure and can be rapidly lifted in case of a film tear. The rollers should be set for maximum pressures that will not cause shifting of the fibers or squeeze-out of resin paste. Pressures at both ends of the rollers should be equal.

Take-up Section

Most SMC machines have a dual turret take-up system for continuous operation. The take-up rolls may run off the end of the conveyor at the same level, as shown in Figure 3.7-1, for horizontal take-up or may be above the end of the conveyor for vertical take-up. The latter setup requires less space.

When one take-up roll is filled, the sheet is cut and transferred to the second roll. Each completed roll is taped to prevent unwinding, and a vapor barrier sleeve is applied. The sleeve may be any material that will contain the styrene monomer within the SMC and prevent ultraviolet light or moisture absorption. Aluminum foil or a monomer resistant film such as nylon are commonly used as sleeve materials.

Standard SMC rolls vary in weight, but about 180 kg (400 lb) is considered maximum for stability or for shipping. Where SMC rolls are stored in-house for subsequent molding, they can be suspended on racks that are either mounted on wheels or designed for transport by forklift vehicles.

Because larger molded parts are becoming more common, bulk packaging concepts that will prevent constant changing of the regular sized rolls at the molding area have now been developed. In one concept, a rectangular box is located at the take-up section of the machine and is filled automatically by equipment designed to distribute the continuous sheet evenly in the form of plies or layers throughout the container. Such containers can hold 900 to 1,350 kg (2,000 to 3,000 lb) of SMC.

Maturation

Sheet molding compound needs to thicken between processing and molding in order to provide a tack free molding compound that releases cleanly from the carrier film. This release normally takes place in a temperature-controlled "maturation" room. Viscosity of the resin paste depends on the resin formulation. Levels between 10 and 100 million centipoise are usually chosen for molding. Maturation room temperature is normally maintained at about 29 to 32°C (85 to 90°F). Depending on the resin formulation, storage time will vary from one to seven days, with most common formulations requiring about three days.

The need for maturation requires interruption of the automatic operation in producing SMC parts and provision of a surge area for inventory of the maturing product. One answer to the problem is to formulate materials that will mature more rapidly. Some work has been done with isocyanates to replace the magnesium compounds presently used as thickeners [5]. Although much faster cure has been obtained, the fastest maturation time is still about four hours – not adequate for in-line processing. This area will require considerable development.

SMC Process Variations

SMC-C and SMC-C/R. A typical setup for processing continuous fiber reinforcements into sheet molding compound is quite similar to that for SMC-R, as shown in Figure 3.7-2. A station is added to the line to feed the continuous strands, after the chopped roving has been deposited onto the moving carrier film [4].

When continuous bundles of roving are used, they will twist when pulled from the inside of the package. This twisting can be minimized by unrolling a conventional low-yield roving from the outside of the package. To insure even distribution of the rovings across the

SMC-C/R MACHINE

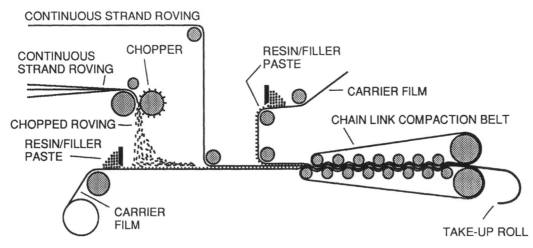

FIGURE 3.7-2. Processing equipment for SMC containing random chopped strands and continuous reinforcements.

sheet width, the continuous roving station must include a roving spacing device. Such a device will normally permit controlled variations in the reinforcement band width.

An alternative approach can be used off-line for the preparation of an SMC compound with unidirectional reinforcement having lengths ranging from 8 to 20 cm (3 to 8 in). This compound has been given the nomenclature of SMC/D. Basically, an SMC-C/R material is produced, as described previously. This material contains continuous fibers on one side of the SMC sheet and random fibers on the other. After the resin paste thickens to at least 3 million centipoise viscosity, the sheet is processed through a long fiber cutter, as shown in Figure 3.7-3. The blades cut through the carrier film and then pierce the continuous rovings into the predetermined lengths.

TMC. Thick Molding Compound (TMC) can be used as an input compound for injection molding as well as for compression molding [6]. Typical processing equipment for TMC is shown in Figure 3.7-4. In this process, the resin paste is fed by a metering pump to the surface of the rolls. Fibers are chopped from continuous strands by a standard chopper/assembly and fall into a hopper that feeds them into the nip of impregnating rolls. The fiber and resin paste feed rates are both controlled to provide the desired fiber to resin paste ratio.

As the fiber/resin paste matrix leaves the nip of the rolls, it is removed by high speed rotating wiper rolls that throw it directly onto the moving carrier film. After an upper carrier film is fed in, the sandwich is carried through a compaction area for formation into sheet. Thickness of the sheet is determined by the speed of this conveyor. A low film speed gives more weight of compound per unit area. Other processing conditions

SMC-D/R
OFF-LINE PRODUCTION PROCESS

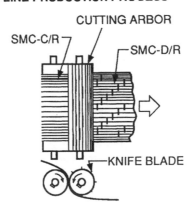

FIGURE 3.7-3. Long fiber cutter for converting SMC-C/R to SMC-D/R.

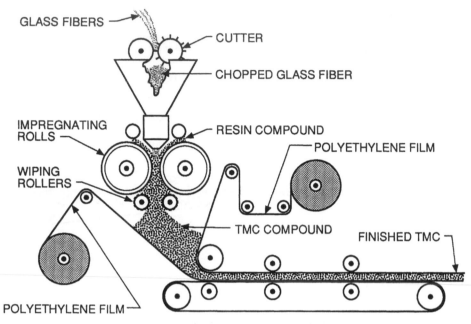

FIGURE 3.7-4. Processing equipment for producing thick molding compound (TMC) [4].

such as resin paste mixing and sheet take-up, for example, are quite similar to those for SMC.

Quality Control

All molding compounds require close quality control of the ingredients in the formulation and of the processing steps during manufacture. A number of important checks can be used to ensure reproducible material with consistent performance [7].

A listing of raw material quality control tests for each SMC ingredient is shown in Table 3.7-1. Some of the tests must be conducted on a regular basis; others are performed on an audit basis. Quality control test methods may be designated by ASTM or SPI standards, although many of the procedures used are developments of the raw material suppliers.

During processing, such variables as environmental controls, resin paste mixing times and amounts, paste viscosities, glass contents, and machine settings must be closely checked or monitored to ensure the production of consistently acceptable sheet molding compound.

Temperature and humidity control is recommended for the SMC processing area. A room temperature of about 24°C (75°F) is best. Humidity is more critical since it directly affects thickening rates of the resin paste and the amount of static electricity generated in the glass-fiber chopping operation. Excessive moisture can upset the controlled water content of the resin mixture, resulting in sporadic thickening, usually at faster than desired rates. Very low humidity contributes to static problems at the machine chopper station. Static buildup can cause poor "lay down" of the glass-fiber blanket and variations in glass content in the SMC roll. For best results, a relative humidity range of 45 to 55% is recommended.

Mixing procedures must be established to assure that all ingredients are added in proper sequence and in the correct amounts and are mixed for the proper time cycles. A continuous mixing system ensures consistent blending of materials while delivering reproducible and controlled amounts of thickened resin paste to the SMC machine.

Resin paste viscosity measurements are essential for the processor and, ultimately, for the molder. Initial viscosity checks are made on resin paste samples collected at the metering blade section of the SMC machine. Initial viscosities should be low to assure adequate wet-out of the glass-fiber reinforcement during processing. Paste samples are retained with the packaged SMC and used for subsequent testing to determine "long-term" viscosities. These readings indicate how soon the SMC is moldable (usually three to seven days) and the length of time the material remains in a processable state.

Monitoring of SMC machine settings is an important process control measure. Conveyor belt speed and glass chopper speed are two process variables that should be checked frequently during production runs. Accurate glass content in the SMC and even distribution across the width of the product are critical processing parameters that can be checked easily and quickly at the machine site.

BMC Compounding

Bulk molding compounds are normally produced using a batch process. The process involves specialized mixing techniques for preparing an easily handled putty-like molding compound containing all the ingredients of the composite formulation. The mixing process usually consists of two separate mixing stations. One prepares the resin paste; the other mixes the glass fibers with the paste. These operations are kept separated because of the need to thoroughly wet and disperse the fillers in the resin and to wet the glass fibers without extensive degradation.

The initial resin paste is prepared in a high speed, high shear mixer. To this mixer are added the resin system, initiator, mold release, and fillers. The high speed disperser mixer is run until the ingredients are thoroughly blended. Excessive temperature can accelerate the thickening mechanism with MgO dispersions and shorten the shelf life of the compound. Therefore, the temperature rise due to mechanical energy input must be closely watched, and the final temperature should be kept to about 35°C (95°F).

The resin paste is then transferred to a second batch mixer that has a sigma blade, a spiral blade, or a combination of blades and single or twin screws. Fillers may be added here also if the paste is not completed in the high shear disperser mixer.

Glass fibers for BMC are normally added at this point in the form of 13 mm (½ in) chopped strands.

Longer fibers tend to be broken by the mixing action. The chopped strands are added at a rapid rate and mixed until they are thoroughly wet with resin. This final mixing time is usually in the order of three minutes.

The completed mix is then packaged in vapor barrier containers either for maturation of the resin paste or for transport to the molding press. Alternatively, the completed compound may be extruded in log form for easier weighing at the press. As with the final mixer, care must be taken not to further degrade the glass in the compound with the extruder.

3.7.4 Molding Technology

After processing, parts from SMC and BMC are formed by compression molding. A measured charge of compound is placed between the halves of a split mold.

Table 3.7-1. Important quality control tests for SMC raw materials.

1. *Polyester Resin* Viscosity Specific Gravity SPI Gel Time Acid Number Water Content Hydroxyl Number Color Thickening Rate	5. *Initiator* Active Oxygen SPI Gel Time (In Control Resin) Infrared (Against Control) Purity Color
2. *Glass-Fiber Reinforcement* Yield Moisture Content Abrasion Resistance Solids Ribbonization (Strand Bond) Stiffness	6. *Mold Release* Moisture Content Apparent Density Particle Size Distribution Melting Point 7. *Pigment* Solids Content Particle Size Distribution SPI Gel Time Infrared Tint and Shade
3. *Filler* Particle Size Oil Absorption Moisture Content Specific Gravity Dispersion	8. *Flame Retardant* Moisture Content Oil Absorption Loss on Ignition
4. *Thickener* Moisture Content Particle Size Purity	9. *Thermoplastic Resin Additive* Solids (Percent of) Specific Gravity Viscosity

Table 3.7-2. General specifications for SMC compression presses.

1. *Type Design:* Open housing, tie rod or frame type, down-acting with self-containing hydraulics.

2. *Guide System:* For open housing type use adjustable, four-point tapered beveled gibs with phenolic laminate wear plates and hardened steel wear strips. For tie rod type, use replaceable bronze full-barrel guides or half-round adjustable tapered gibs.

3. *Platen Area:* Area for mounting die set. Specified on basis of molder's needs.

4. *Bolster:* The press bed plate should have provisions for cooling.

5. *Daylight:* Full open distance between bolster and top platen. This can be established on the basis of the depth of the part.

6. *Stroke:* Distance the top platen is capable of moving. Can be established by multiplying the depth of the deepest anticipated molded part by 3.

7. *Shut Height:* Distance between bolster and top platen when press is at the end of its stroke. It is seldom practical to require less than 30 cm (12 in) of shut height.

8. *Force:* Adjustable from either 10% of maximum or 50 tons, whichever is lower, to maximum tonnage within 5% accuracy.

9. *Deflection:* Platen-bolster bending deflection. Not to exceed 0.02 mm (1 mil) per 30 cm (12 in) along the diagonal between the housings when uniformly loaded over ⅔ of the platen area and submitted to the full force.

10. *Platen Parallelism:* When a 12.5 cm (5 in) diameter cylinder is eccentrically supported 60 cm (24 in) off the centerline by the crosshead dead weight, the platen parallelism should be within 0.1 mm per 30 cm (4 mils per foot).

11. *Rapid Advance:* 15 m/min (600 in/min) or faster.

12. *Intermediate Advance:* Adjustable to full bypass. Rates determined by purchaser.

13. *Final Pressing Speed:* Adjustable, 0 to 37.5 cm/min (0 to 15 in/min).

14. *Break Away:* Pull back capacity at 20 to 25 percent of rated force tonnage.

15. *Rapid Return:* 15 m/min (600 in/min).

16. *Pressure Buildup:* Five seconds maximum, from zero to maximum tonnage.

Table 3.7-3. General specifications for SMC molds.

1. *Mold Block Material:* P-20 steel forgings for surface appearance of high production parts. Ribs and slots can be cut using an Electrical Discharge Machine (EDM).

2. *Mold Block Thickness:* The mold must have sufficient beam strength to resist deflection from molding pressures. This should be calculated for each mold. Assuming the mold compound acts like a thick liquid, the press platen rigidity should be considered in the calculation.

3. *Mold Parallelism:* Within 0.13 mm (5 mils) at any point between the two outside mounting surfaces.

4. *Part Cavity Thickness Variation:* Within 0.25 mm (10 mils) of any one surface.

5. *Shear Edge Travel (Telescope):* 3.1 mm (⅛ in) for simple parts. 3.1 to 6.2 mm (⅛ to ¼ in) for complex parts. 3.1 to 37.5 mm (⅛ to 1½ in) for parts requiring special charge patterns.

6. *Shear Edge Clearance:* 0.05 to 0.1 mm (2 to 4 mils).

7. *Shear Edge Draft:* One degree in cavity shear edges parallel to mold movement. For textured surface, cavity draft must be increased one degree for every 0.025 mm (1 mil) of texture depth.

8. *Shear Edge Hardness:* 50 to 55 Rockwell C for a depth of 1.6 mm (1/16 in).

9. *Guide Pins and Bushings:* Minimum diameter 2% of length plus width of die. Bushings should have air vents and grease fittings.

10. *Shoe Blocks and Wear Plates:* Shoe blocks should be integral with cavity or keyed and bolted. Wear plates should be hard bronze. Note: Shoe blocks are required when lateral loads will cause deflection of the guide pins. They can be positioned either at the four corners of the mold or at the centerline edges.

11. *Surface Finish:* Appearance parts: Cavity—400 grit long stroke—unbuffed; core—320 grit unbuffed. Non-appearance parts: cavity—320 grit buffed; core—320 grit unbuffed.

12. *Chrome Plating:* 0.01 to 0.05 mm (0.5 to 2 mils) hard chrome, blemish free, and buffed to a high luster.

13. *Temperature Variation:* Within 2°C (5°F) on die surface.

14. *Positive Stops:* Split with beveled edges. Half on cavity and half on core with 0.38 mm (15 mils) shims removable for positive pressure molding.

15. *Shipping Straps and Stop Spacers:* Shear edges must be disengaged during shipment using four 10 cm × 2.5 cm × 1.3 cm (4 in × 1 in × ½ in) steel straps bolted between cavity and core with spacers between the stops.

16. *Eye Bolt Holes:* Four tapped holes in each half to allow lifting from either surface.

The heated mold is closed, pressure is applied so that the compound flows to fill the mold cavity, and the compound is allowed to cure before removal from the press. The molding conditions—press speed, temperature, pressure, and cure time—depend on the complexity of the part, its shape and thickness, and the type of compound.

Press Specifications and Design

Molding of SMC parts has required certain adaptations and improvements in the hydraulically driven compression presses used. Table 3.7-2 presents important requirements and specifications of these presses, some of which are discussed in greater detail below.

Compression Pressures. The type of compound to be molded and the mold configuration are the primary factors determining pressure requirements of a press. Generally, a range of 3.4 to 13.8 MPa (500 to 2,000 psi) is adequate. Low shrink, low profile compounds, with high volume fractions of filler and glass, require high pressures if the molded parts are to have top quality surfaces. High pressures are also necessary with more complex parts—those that have ribs and bosses or that require a deep draw. As compression molding has progressed to larger parts using low profile compounds, the need for high tonnage presses has grown.

Press Speeds. A compression molding press must have a carefully controlled closing cycle. In the molding process, the moveable upper platen advances rapidly to within centimeters of the closed position. Then the hydraulic system switches to a lower speed, requiring several seconds for the mold to come to a fully closed position. This type of action permits the compound to flow and fill out the mold cavity before full pressure is applied. If the press closes too slowly, the compound may begin to cure before the mold is completely filled.

Guide Systems. The guide system for a compression press controls the position of the top platen as it moves to engage the guide pins and shoe blocks of the mold. Over the past few years, press designs have evolved from one in which resistance to lateral forces was borne almost wholly by the press guides to one in which mold shoe blocks are used to carry the predominant lateral load. In either case, the total guiding system must be designed and constructed to prevent damage and excessive wear of the mold shear edge.

Guide mechanisms may include tapered parallel gib systems that can be adjusted to close clearances. When such systems are used, the gibbing wear surfaces should be lined with self-lubricating synthetic materials to eliminate the possibility of lubricant contamination around the mold area.

Platen Parallelism. It is extremely important that the press platens remain parallel during molding. Otherwise, the molded parts will have thickness variations across the surface in proportion to the final platen position. Thickness variations can cause nonuniform cure rates that may create internal stresses causing the part to become wavy. In addition, nonparallel platens can cause part of the charge to reflow after it has partially set, resulting in fiber separation, flow lines, and reduced physical properties.

Parallelism is very difficult to maintain when high eccentric loads are present. One approach is to provide rigid guiding as part fillout is approached. At part fillout, full tonnage retractable stops force the upper platen into parallelism with the press bed. The stops are then decompressed, allowing the molded part to cure under full pressure.

In another approach, linear variable displacement transformers (LVDTs) monitor the position of each corner of the press ram. Servo-controlled hydraulic cylinders at each corner then activate differentially on signals from the monitor, providing a force network that maintains parallelism [5].

Mold Specifications and Design

Because compression molding normally involves long production runs at high pressures, there is a need for strong, accurate molds. General specifications for molds used in producing compression molded parts are shown in Table 3.7-3.

To meet the demands of high volume production and good appearance, molds must be of durable material, such as prehardened steel. For high quality surface finish, P-20 steel is recommended. When glossy, smooth surfaces with little or no post-molding treatment are needed, the molds should be chrome plated, buffed to a high luster, and blemish free.

There are several heating methods that can be used with the metal tooling. Steam heating is the least expensive and is usually recommended. Alternatives are electric and oil heating.

To facilitate part removal, ejection pins are used with the tooling. Pin positions are highly dependent on tool and part design.

Compression Molding Process Control

There are several processing variables that affect the quality and the reproducibility of molded parts. However, no single set of manufacturing conditions will meet the demands of the many formulations and tooling designs being used. Variables that the molder must consider include mold temperature, mold pressure, charge pattern and placement in the mold, press closing speed, and curing time. Analysis of these variables, frequently accompanied by some trial and error experiments, is needed to arrive at the best processing conditions for a particular application.

Molding Temperature

Typical SMC and BMC formulations are cured at nominal temperatures of about 149°C (300°F). Temperature variations of a few degrees can result in premature gelling in some areas of the mold if it is too hot or longer cure times than expected if some areas of the mold are too cold. Temperatures as low as 132°C (270°F) or as high as 160°C (320°F) are possible with the proper formulation, but this range of temperatures would never be tolerated within a single mold cavity.

Temperature control is extremely important in the molding process. The objective is to obtain the fastest possible cure at settings that are easy to maintain, without any degradation of the material, and with near uniform heating throughout the charge material. Accurate temperature control will improve productivity, reduce internal stresses in the molded part, and help ensure reproducible physical properties. The biggest challenge is to obtain an appropriate temperature distribution across the entire mold. In the molding process, the charge is placed at room temperature on the heated mold. As the press is closed, the center of the mold cools somewhat while the heated compound extrudes toward the perimeter. When production rates are such that the total molding cycle is about five minutes or longer, the mold heating system will usually heat the cooler center back to the proper temperature. Under conditions where a faster rate of production is desired and the total cycle time approaches two minutes, the slightly cooler center can govern the overall curing time in the mold.

This problem can be overcome with proper thermal control by analyzing heat flow in different zones, calculating heat transfer between these zones, and then varying mold heating and cooling line locations appropriately. In this case, the thermal energy input to the mold is balanced to the actual molding requirements [5].

Charge Pattern and Placement. The charge pattern for a mold should be as simple as possible. Rectangular plies that give 30 to 50% flow in the mold are preferred.

Consistent placement of the charge in the mold assures reproducible flow time to all parts of the mold and is also important for proper temperature distribution. Charge placement also affects glass-fiber orientation, a condition which is caused by flow in the mold and which can result in variation in part physical properties. Obstructions in some mold designs cause interruption and subsequent rejoining of the flowing material. This interruption can cause some local weaknesses, referred to as *knit lines*. Proper charge placement is a key to minimizing or eliminating these conditions.

Various automatic loading and unloading devices have been developed for compression presses. Common faults of these devices have been that they are too slow, that they pierce the part causing resin-rich spots to develop during molding, and that they position the charge erratically. However, these faults have been overcome in more recent devices. With accurate loading in less than five seconds by devices that grip, rather than pierce, the charge is possible. One design, for example, employs an automatic loader-unloader, which can be built for a particular press, and uses interchangeable tooling to adapt its functions to any mold used in that press [8].

Molding Pressure. Parameters to consider in establishing molding pressures are viscosity of the resin paste, part design, and, to some extent, quality of the part surfaces. Low shrink, low profile compounds require relatively high viscosities to provide enough flow of the compound in the mold. This requirement is particularly true with large parts. With relatively large molds and with designs requiring deep draws, higher pressures are required (up to 13.8 MPa or 2,000 psi). Satisfactory flow and fillout will be obtained at slightly lower pressures for parts with flat surfaces.

Press Closing Speed. Selection of press closure rate depends primarily on mold temperature and gel time. Higher molding temperatures and faster curing formulations require faster press closure for rapid pressure build in the mold. If closing is too slow, material will pre-gel in the mold. If it is too fast, resin wash will result. Normal closing time will range from one to ten seconds.

Cure Time. The primary factor controlling cure time is the resin initiator system, followed closely by mold temperature and part thickness. Depending on these factors, cure times range from one to five minutes. There are two basic methods for determining cure time.

The first method is a simple observation of the molded part. There is no technical problem associated with too long a cure cycle. However, as the cure time is reduced, undercure can be observed either by internal blisters in thicker sections such as bosses or by distortion and loss of gloss in the part as it continues to cure after removal from the press.

The second method is an instrumented technique that provides a thermal profile of the material throughout a cure cycle. Thermocouples extended into the tool cavity can be used to monitor temperatures of the molding compound during heat-up to peak exotherm and subsequent cooling to mold temperature. The cure time is defined as the time required for the material to advance through the heat-up stage to a point just beyond peak exotherm. Normally, the thermocouple is positioned to monitor the temperature in the thickest section of the molded part.

The cure time is presently a limiting factor with respect to increased production rates. The actual molding of the part occupies only about 10% of total press cycle time; the remaining 90% is cure time. Several means to separate the curing function from the pressing function have been suggested. One approach would be a shuttle mechanism to alternate two molds in and out of the press [5]. Another concept is multi-station transfer lines that would break the molding operation into its functions of loading, applying pressure, and unloading. Molds would be transported from station to station with ample distance between each station to allow for cure.

These are still just concepts and have not yet reached the hardware stage. However, if compression molding is to develop to its full potential and overcome present productivity shortcomings, further development of these or other concepts will be required.

Quality Control

There are a number of quality control checks recommended for the molder who does not manufacture his own SMC. These incoming material tests are used as initial checks on the acceptability of the shipment received from the SMC processor. Generally, the molder will establish a number of specifications that the SMC

Table 3.7-4. Important quality checks on SMC before molding.

1. *Sheet Weight:* The weight per sheet area should be within the manufacturer's specification.

2. *Glass Content:* The glass content of the SMC should be within the manufacturer's specification.

3. *Film Release:* The carrier film should release from the product easily and with a minimum of residue.

4. *Tackiness:* The product should not be overly tacky, to the extent that it would present handling or mold loading problems.

5. *Wet-out:* The product should be free of dry areas of glass reinforcement.

6. *Color:* The product should have good color dispersion (where colorants are specified). Poor color dispersion usually indicates inadequate mixing of the resin paste.

7. *Contamination:* The product should be free of excess foreign matter.

8. *Viscosity:* The product should be chemically thickened to a viscosity level that will produce acceptable parts when molded in a production tool.

9. *Cure:* The product should contain sufficient quantities of the initiator to ensure satisfactory cross-linking of the polyester resin formulation when molded in the production tool.

must meet to assure consistent quality of its handling and molding characteristics from shipment to shipment. Table 3.7-4 describes basic requirements that should be checked on each material lot to verify the quality of the shipment.

3.7.5 Mechanical Properties

One interesting characteristic of composites is that the material itself and the molded part are created at the same time. This process is unlike fabricated metal parts for which the properties of the raw stock are measurable and, also unlike molding thermoplastic parts, for which the physical state simply is changed from a solid to a liquid and back again.

One principle to keep in mind when evaluating or reviewing published data on the performance of BMC or SMC composites is that the mechanical strength generally depends on the combined effect of the amount of reinforcement, the length of the reinforcement, and the arrangement or orientation of the reinforcement in the finished part.

The basic strength of BMC and SMC composites is directly related to the amount of glass fiber in the molded part, as shown graphically in Figure 3.7-5 for

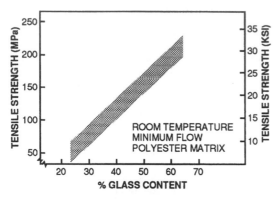

FIGURE 3.7-5. Tensile strength of SMC composites with random chopped strands.

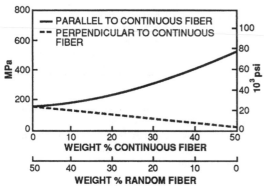

FIGURE 3.7-6. Tensile strength of SMC-C and SMC-C/R at 50% fiber volume fraction.

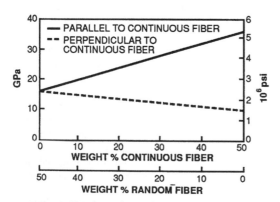

FIGURE 3.7-7. Tensile modulus of SMC-C and SMC-C/R at 50% fiber volume fraction.

Table 3.7-5. Basic mechanical properties of SMC and BMC compression molded composites.

	SI Units			English Units		
	SMC	BMC	Units	SMC	BMC	Units
Glass Fiber Content, by wgt	15–30	15–35	%	15–30	15–35	%
Tensile Strength	55–140	27–70	MPa	8–20	4–10	ksi
Tensile Modulus	11–17	11–17	GPa	1.6–2.5	1.6–2.5	Msi
Ultimate Tensile Elongation	0.3–1.5	0.3–0.5	%	0.3–1.5	0.3–0.5	%
Compressive Strength	100–200	140–210	MPa	15–30	20–30	ksi
Flexural Strength	125–205	70–138	MPa	18–30	10–20	ksi
Flexural Modulus	10–14	10–14	GPa	1.4–2.0	1.4–2.0	Msi

the tensile strength of an SMC composite over the range of glass contents, which can easily be produced in these compounds.

The effect of the length of the reinforcement can be seen by comparing the mechanical properties of BMC with SMC. Normally, BMC contains 13 mm (½ in) fibers that have been intensively mixed with the resin paste and actually may be shorter than the original length. SMC, on the other hand, normally contains 25 mm to 50 mm (1 to 2 in) fibers that are not shortened by the SMC machine and are rarely degraded in molding. This comparison is shown in Table 3.7-5.

The third element of the condition of the reinforcement is fiber orientation. This element is not easily predicted for chopped fiber reinforced composites, but its effect can be measured if mechanical properties are obtained from molded parts of different shapes or from flat composites in which specific changes in the flow of the compound were introduced in the molding operation.

Another technique for documenting the effect of fiber orientation is to examine the properties of press molded composites in which the overall orientation has been controlled by the introduction of unidirectional fibers. Some data for flat composites molded from combinations of continuous strand SMC and random chopped strand SMC are shown in Figures 3.7-6 and 3.7-7. These figures for composites at 50% glass-fiber volume fraction show that the strength and modulus in the principal directions change dramatically as the amount of reinforcement oriented in the direction of the load also changes.

Influence of Process and Design

The compounding process for BMC and SMC establishes the nominal amount and length of glass fiber; there is not much the molder can do to change these values.

However, the third variable, orientation of the fibers, can be influenced both by the molder and by the designer of the part. The left side of the X-axis in Figures 3.7-6 and 3.7-7 represents an SMC composite in which the reinforcement is totally random. Predominant flow in one direction can orient the fibers of a chopped strand reinforced SMC in the direction of flow, giving the composite a "directional" characteristic. In the extreme, a chopped fiber reinforced SMC will achieve properties at the far right side of these figures. In practice, the tensile strength in the principal directions can achieve ratios of up to three to one if unidirectional flow occurs for a significant distance [9,10].

Therein lies the enigma of published engineering data for any press molded composite system. Normally, published data on BMC and SMC are measurements that have been carefully documented from specimens cut out of flat sheets of the molded composite. The thickness of the sheet is usually controlled to a level between 2 and 6 mm (0.08 and 0.25 in) that provides specimens for testing by established ASTM test methods. Actual molded parts are rarely flat and often contain designed thickness variations that exceed the values of tested specimens. In addition, each part will provide a slightly different flow situation to the molding compound that can greatly influence the orientation of the reinforcement.

Thus, both the designer and the fabricator of a compression molded part have limited control over the physical and mechanical properties of the molded composite part [11].

3.7.6 References

1. ENGLISH, L. K. "Fabricating the Future with Composite Materials, Part I: The Basics," *Materials Engineering*, 4:15 (September 1987).

2. *Fiberglas-Plastic Applications in Appliances and Equipment*, Owens-Corning Fiberglas Corporation, Pub. No. 5-EA-5754 (1972).

3. McCLUSKEY, J. J. and F. W. Doherty. "Sheet Molding Compounds," *Engineered Materials Handbook, Vol. 1, Composites*, ASM, Intl., Metals Park, OH, p. 157 (1987).

4. *Structural SMC: Material, Process, and Performance Review*, Owens-Corning Fiberglas Corporation, Pub. No. 5-TM-8364 (October 1978).

5. BLATT, R. W. "Improving the Efficiency of SMC Equipment," *Plastics Machinery and Equipment*, 8(1):22–28 (January 1979).

6. MEKJIAN, A. "Injection Molding TMC for Class A Surfaces," *Plastics Engineering*, 35(9):56–58 (September 1979).

7. *Sheet Molding Compound*, Owens-Corning Fiberglas Corporation, Pub. No. 5-TM-6991-A (June 1976).

8. "Automatically Load/Unload SMC Molding Presses," *Plastics Technology*, 24(12):131–135 (November 1978).

9. HERMAN, E. A. "How to Meet Automotive Standards with SMC," *Plastics Engineering*, 36(10):31–33 (October 1980).

10. DENTON, D. L. "Effects of Processing Variables on the Mechanical Properties of Structural SMC-R Composites," *Proceedings of the 36th Annual Technical Conference of the Reinforced Plastics/Composite Institute*, Washington, D.C., Section 16-B (February 1981).

11. HIRAI, T. "Design Concept of SMC Molding to Prevent a Fault Caused by Flow State," *Proc. ICCM & ECCM, Vol. 1*, F. L. Matthews et al., eds., Elsevier Applied Sci., London, p. 1.121 (1987).

SECTION 3.8

Injection Molding

3.8.1 General Principles

Injection molding is a high-pressure process that was developed initially for molding thermoplastics but has been adapted for conventional thermosets and even bulk molding compounds (BMC) [1]. In this process the molten (plasticized) charge is forced under pressure into the cavity of a closed die, where the charge is formed in the shape of the cavity.

In thermoplastics technology the charge is plasticized (melted) by a combination of heat and mechanical action, and the molten charge is injected into a relatively cold mold. The charge solidifies in the mold and, when sufficiently advanced, the mold may be opened and the part ejected.

For conventional thermosets which are solid at ambient temperatures, the plasticization stage is similar, but the charge is injected into a hot mold where it cures. When the cure is complete (or at least sufficiently advanced to give some rigidity to the part), the mold is opened and the part ejected hot. This is the fundamental difference between the two variants of the process. When molding thermosets, it is necessary to exercise fine control over the temperature of the charge prior to injection to avoid premature cure.

BMC is usually supplied as a viscous paste that may be further plasticized by heating, as is usually the case, or may be conveyed essentially at ambient temperature and injected into the hot mold under pressure.

Although the processes are superficially similar for all three basic types of material, the regions calling for critical control differ considerably. Likewise, equipment must usually be purposely made for one class of materials, and it is seldom possible to utilize a single molding machine for more than one type of material.

The molder will be concerned with the soundness, dimensional accuracy and stability, surface finish, mechanical and physical properties, and, above all, the overall consistency of the moldings. The manufacturer will also be concerned with the speed of production, the degree of critical control required and the cost of achieving it, machine reliability, material wastage, mold life, and the overall production reliability as reflected by the rate of production of acceptable moldings. The overall economics will be determined by the specification of the feedstock, the levels set for production acceptance parameters, the cost of materials, labor, energy, and capital and maintenance costs of machines and molds. In practice, many of these factors are interlinked so that materials and process selection is no trivial matter.

3.8.2 The Principles of Operation of the Injection Molding Machine

There are many variants of the basic injection molding machine, but all may be broken down into a number of subunits:

1. Raw material feed section
2. Plasticization section
3. Compression and injection section
4. Mold closing and ejection control section

To this list may be added devices for delivering feedstock to the machine and removing finished moldings and waste material (see Figures 3.8-1, 3.8-2, and 3.8-3). For thermoplastics and conventional thermosets, where the feedstock is solid—in the form of powder, granules, or pellets—the feed section may be a simple conical hopper, charged manually or by pneumatic conveyor. The contents are often subjected to vibration or agitation to prevent clogging.

The plasticization section in modern machines is now almost universally a screw pump, although plunger action machines are still used by some specialized molders. The screw pump serves to convey the solid

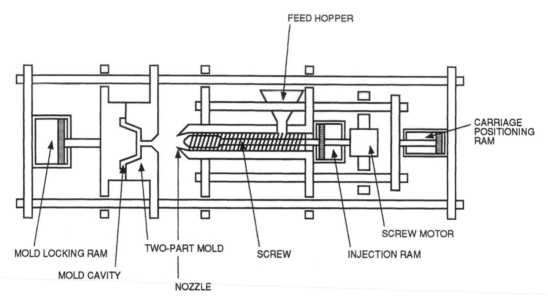

FIGURE 3.8-1. Screw pre-plasticizing injection molding machine.

feedstock from the hopper through a heated barrel section where the charge is melted. The action of the screw in the barrel induces a strong stirring action in the charge, which ensures an efficient transfer of heat from the hot barrel walls to the charge. In addition, further heat may be generated by the mechanical action of the screw on the charge. Whether this effect can be exploited is dependent upon the nature of the charge. This effect, in turn, influences the screw design requirements and the machine operating characteristics. It should be noted at this stage that the plasticization action is part thermal and part mechanical and that the latter action strongly influences the homogeneity of the plasticized charge. The process is much more complex than simple melting.

The plasticized charge is carried forward by the screw and builds up pressure against the shut-off valve at the front end of the barrel. This pressure causes the screw to move back against a controlled back pressure in the main hydraulic cylinder so that a reservoir of plasticized material is formed ahead of the screw. When sufficient charge has been plasticized, the screw rotation is stopped. The carriage is advanced so that the nozzle is pressed against the mold, the nozzle shut-off valve is opened, and the main hydraulic ram is actuated to drive the screw down the barrel, thus injecting the plasticized charge into the mold. The ram pressure is

maintained while the charge starts to solidify. Then the nozzle shut-off valve is closed, the carriage retracted,[5] and the screw restarted to plasticize the next charge. At the appropriate time, the mold is opened, the molding, sprue, and runner ejected, and the mold closed again to await the next injection cycle.

In a plunger machine (Figure 3.8-3) the sequence is similar except that the plunger merely reciprocates. Plasticization is accomplished by forcing the charge through the narrow annulus between the heated barrel and the "torpedo." On the forward stroke the material at the front of the barrel is injected while fresh material is forced past the torpedo for the next charge. As the plunger is retracted past the hopper part, a fresh charge is collected. Several "shots" may be resident in the barrel at any one time. The residence time is set to ensure adequate plasticization, which in this case is accomplished solely by heat transfer from the barrel, although machines with heated torpedos have been made. The plunger machine is inherently simpler, but the screw is a much more efficient plasticization device.

[5]Carriage retraction that ensures a positive sprue break action is often the best practice for FRTP. Other materials may be molded with the carriage permanently forward, whereas others utilize "hot runner" systems. For some materials, open nozzles are preferred (over positive shut-off valves) in order to restrict the range of back pressures available.

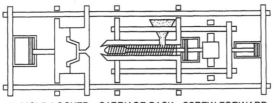

1. MOLD LOCKED—CARRIAGE BACK—SCREW FORWARD

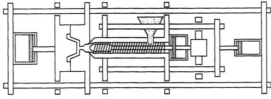

2. CARRIAGE FORWARD—SCREW ROTATION STARTS

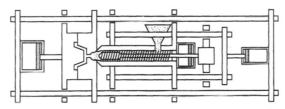

3. PLASTICIZATION COMPLETE—SCREW BACK—ROTATION STOPS

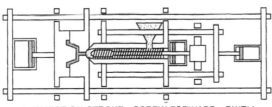

4. INJECTION STROKE—SCREW FORWARD—DWELL

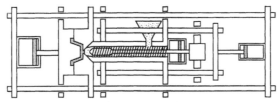

5. PLASTICIZATION COMPLETE—SCREW BACK—ROTATION STOPS

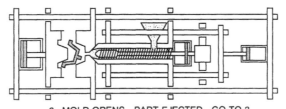

6. MOLD OPENS—PART EJECTED—GO TO 3

FIGURE 3.8-2. Operating cycle of injection molding machine.

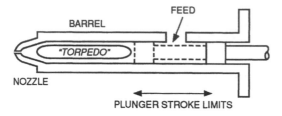

TORPEDO SPLITS FLOW INTO ANNULUS BETWEEN
HEATED BARREL WALLS AND CENTRALLY
SUPPORTED TORPEDO BODY.

FIGURE 3.8-3. Plunger type injection molding machine.

3.8.3 Design and Control of Injection Molding Machines

The principal design parameters are as follows:

1. Mold closing (locking) force
2. Plasticization rate
3. Injection pressure
4. Maximum shot weight
5. Dry cycle time

The mold locking force is the main factor that determines the maximum molding dimensions and the overall mass of the machine. The required locking force must be greater than the product of the injection pressure and the projected area of the molding at the mold parting line. Since injection pressures (for thermoplastics) are sometimes as high as 200 MPa (30,000 psi), it follows that locking forces must be around 20,000 metric tons force for each square meter of projected area or 1–2 metric tons per cm² (approximately 5 tons/in²). In practice, values two to three times as high as those indicated from the pressure × area calculation are often employed to ensure against "flashing." Currently, the largest machines in common usage have locking forces of 2,000–5,000 metric tons force, i.e., 20–50 MN. Moldings of over 1 m² projected area can be made but at much lower pressures.

The plasticization rate is determined by the barrel diameter and length and by the design of the screw. Within given limits of length and diameter, the plasticization rate is determined by the effective heat transfer from barrel to charge. The temperature differential (from ambient to injection temperature) required for a

given material is critical, as is its ability to withstand overheating. Thus, by raising the barrel temperature, a faster heat transfer rate may be achieved; however, this is acceptable practice only if no degradation occurs through local overheating of the material in contact with the barrel wall. The effective mixing action is determined by the screw design. A finer screw pitch gives better mixing but a slower feed rate for a given power input. When the feedstock is capable of accepting mechanical shearing before melting, a fine pitch screw and high power input may be used so that the heat transfer from the barrel is augmented by the mechanical shearing work (adiabatic heating). Raising the back pressure increases the work done on the material by the screw and decreases throughput for a given power input.

As shown in Figures 3.8-4a, 3.8-4b, and 3.8-4c, the screw incorporates a compression stage, usually 2:1 to 3:1. Higher compression ratios also result in more work on the charge and greater homogeneity. The plasticization rate is thus determined by a number of complex interactions. In practice, however, the plasticization rate is seldom the rate-limiting parameter in the injection molding process.

The injection pressure is determined solely by the combination of screw (ram) diameter and the capacity of the hydraulic ram, which performs the injection function. Higher pressures call for stronger design, but this factor is generally of less influence than the mold locking force in determining the overall bulk of the machine.

The maximum shot weight is determined simply by the screw diameter and injection stroke. Machines are conventionally rated on their shot weight capacity in polystyrene [e.g., 100 g (4 oz)]. This is, of course, a shot volume, and weights will vary according to the density of the materials molded.

The dry cycle time is the time taken for the machine to go through a complete molding cycle at maximum injection speed and without allowance for "dwell" and "cooling" times. This time may vary from ~5 sec for a small 50 g (2 oz) machine up to perhaps 60 sec for the largest contemporary machines. Real cycle times are considerably longer, and the "cooling in-mold" time is most often the critical factor.

A typical cycle for a 25% (volume) glass-fiber filled polyamide 6.6 (nylon 6.6) molding of 30 g molded on a 50 g shot weight/150 kN locking force machine at 150 MPa injection pressure (2 oz/15 ton/22 × 10³ psi) is given in Figure 3.8-5.

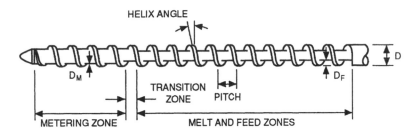

TYPICAL SCREW WITH SHARP TRANSITION

FIGURE 3.8-4a. Typical injection molding machine barrel and screw.

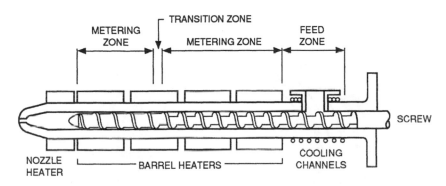

FIGURE 3.8-4b. Typical screw with sharp transition.

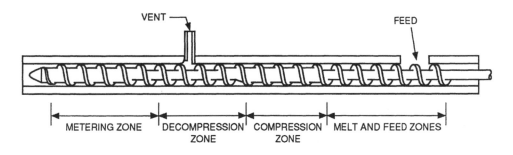

VENTED BARREL WITH TWO-STAGE SCREW

FIGURE 3.8-4c. Vented barrel with two-stage screw.

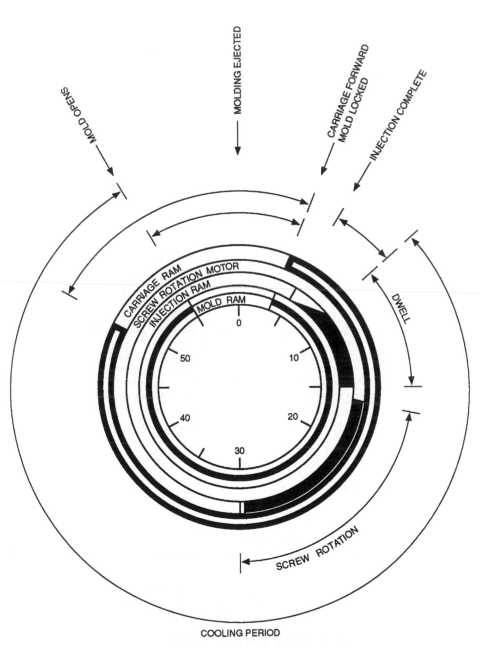

FIGURE 3.8-5. Schematic of a 60 sec injection molding cycle.

The cycle commences in the screw rotation phase, which continues until the correct charge volume has been plasticized. This operation is concurrent with the mold cooling phase of the previous cycle. When the mold has been cleared and locked, the injection phase commences. The injection rate (i.e., the ram speed) is controlled by a restrictor valve at the ram and has considerable influence on mold filling, surface finish, and general quality of the molding, as will be discussed further. At the end of the injection stroke the full injection pressure is built up on the material in the mold, and this pressure is maintained during the "dwell" (also called "screw forward time"–SFT) period. The purpose of the dwell is to allow a measure of "feeding" as the polymer charge shrinks upon solidification in the mold. The useful dwell period is terminated when the polymer in the gate becomes solid. At this point no more material can be forced into the mold, neither can any flow back out, so the pressure may be relaxed. The cooling phase proper then commences, and the shut-off valve may be closed and the next plasticization phase initiated. The molding must then be allowed to cool until the charge has solidified to a stage where the part may be safely ejected without risk of unacceptable distortion. In virtually all practical situations, the cooling phase imposes the ultimate limitation on total cycle time and hence production rate.

In the following subsections the influence of machine settings, with particular regard to fiber reinforced thermoplastics, is discussed in more detail, but it is first appropriate to consider the influence of some fundamental aspects of barrel and screw design. These components combine to determine the plasticization capacity of the machine, and their design also influences the overall energy input requirements and the quality of the molding. This influence is especially true in the case of the more sophisticated "engineering" thermoplastics, which are often fiber reinforced. The length to diameter ratio (L/D) of the screw determines the relative heated surface to which the charge is exposed as it passes through the barrel. The other important feature is the screw compression ratio (C/R) (see Figures 3.8-4a, 3.8-4b, and 3.8-4c). Most thermoplastics can be adequately processed with screws of 10:1 or greater L/D and C/R of 2:1–3:1. Another important design feature is the transition from the feed to the compression zone of the screw. A sharp transition (Figure 3.8-4b) is suitable for crystalline polymers, such as nylon, that exhibit a sharp melting point. The settings should be arranged so that the charge is substantially liquid at the transition. For other polymers with a gradual softening range, a shallow transition may be utilized. This transition allows the plasticization process to be assisted by mechanical work input. Shallow transition zones are commonly employed in extruders, but mechanical plasticization is seldom exploited in injection molding machines since the plasticization stage is not normally critical in determining the production rate. Another design feature of considerable significance is the introduction of the two-stage vented screw into injection machines. This design features a decompression zone between two compression zones (Figure 3.8-4c). Volatiles may be drawn off at this decompression zone. This is of considerable importance when handling materials that are hygroscopic—e.g., polyamides, polycarbonates, etc.—and those subject to thermal degradation induced volatiles. The use of such a screw can reduce the need for pre-drying the charge and often results in a sounder and more consistent product than that produced by the conventional route. Even with such equipment, however, it should be emphasized that some pre-drying of hygroscopic materials is generally advisable unless its moisture content is sufficiently low.

3.8.4 Machine Operating Settings

The general principles of operation of the screw preplasticizing injection machine for thermoplastics molding have been discussed in the previous subsections. In this segment, the influence of the individual machine settings and their separate and combined influence on the process of molding thermoplastics, in general, and fiber reinforced thermoplastics, in particular, are discussed in more detail. The special requirements for molding thermosets and BMC materials are treated in separate discussions.

We will first consider the control of plasticization. This control is accomplished by a combination of heat transfer from the heated barrel by circulation of the charge by the screw, by adiabatic heating resulting from mechanical work done by the screw on the charge, and by the rapid flow induced during the injection phase of the cycle. Control of plasticization is accomplished by the settings of the barrel and nozzle heaters, the screw design, screw rotation speed, back pressure control, nozzle and gate dimensions, and injection speed.

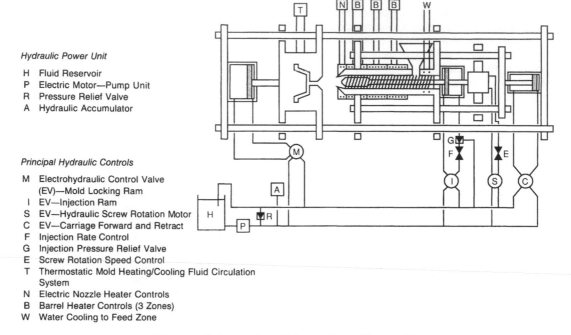

Hydraulic Power Unit

H   Fluid Reservoir
P   Electric Motor—Pump Unit
R   Pressure Relief Valve
A   Hydraulic Accumulator

Principal Hydraulic Controls

M   Electrohydraulic Control Valve
     (EV)—Mold Locking Ram
I   EV—Injection Ram
S   EV—Hydraulic Screw Rotation Motor
C   EV—Carriage Forward and Retract
F   Injection Rate Control
G   Injection Pressure Relief Valve
E   Screw Rotation Speed Control
T   Thermostatic Mold Heating/Cooling Fluid Circulation
     System
N   Electric Nozzle Heater Controls
B   Barrel Heater Controls (3 Zones)
W   Water Cooling to Feed Zone

FIGURE 3.8-6. Basic control circuit for injection molding machine.

The barrel is usually divided into two to four zones for better heat control with separate control of the nozzle heaters (see Figures 3.8-4 and 3.8-6). The solids transport phase of feeding the solid material from the hopper to the melting zone is essentially plug flow aided by a high circumferential friction between the barrel and charge. It is thus advantageous that the barrel be cool in this region. The barrel is often axially grooved on the internal bore to help prevent the charge from rotating with the screw. The barrel is usually water-cooled in this feed region.

The melting zone commences 2–3 screw diameters along the barrel, and the heaters must be set at as high a temperature as is consistent with rapid melting while avoiding burning or degradation. The actual safe temperature depends on the screw rotation speed and the sensitivity of the polymer to overheating. Low thermal conductivity and high melt viscosity of thermoplastics reduces the possibility of convection. This lack of convection increases their susceptibility to local overheating. It should also be noted that the molding cycle involves a relatively long phase when the screw is sta-

tionary; thus, the charge is static and in contact with the hot barrel. Therefore, it is usual practice to increase the barrel temperature progressively from the feed end towards the nozzle. The nozzle temperature is usually set lower than the final barrel zone temperature to prevent burning as the charge is adiabatically heated during the (rapid) injection stroke. If high injection speeds and small nozzle and gate dimensions are used, the tendency to "burn" in this way is increased because of the increase in shear rate experienced by the charge on injection. Likewise, high back pressure and fast screw rotation result in more mechanical work on the charge during plasticization, with better heat transfer between barrel and charge due to the more efficient charge circulation.

In general, these effects can be utilized to increase the plasticization capacity of the barrel/screw unit. However, when operating with fiber reinforced thermoplastics, there are at least two additional considerations— fiber attrition and screw and barrel wear.

During the plasticization and injection phases of the molding cycle, the charge is subjected to high shear rates, and, whereas this assists plasticization of the polymer, it also leads to fiber attrition, whereby the reinforcing fibers are actually broken up into shorter lengths. This effect can seriously impair the reinforcement efficiency. Thus, as a general rule for FRTP, it is advantageous to utilize low screw rotation speed, low injection speed, and low back pressure. These factors reduce fiber attrition and also decrease the rate of wear on the screw, barrel, and mold, which can be rapid unless suitable settings and materials are chosen.[6] For the same reasons, it is better to use generous nozzle and gate apertures.

It is useful to note at this stage that thermoplastics filled with 20–40% fiber or mineral filler require a considerably lower total heat input for plasticization than the corresponding unfilled material. This requirement is due to lower specific heat of the filler and a relatively smaller volume to absorb the heat of melting.

Against this argument, however, must be taken the fact that most thermoplastics exhibit "pseudo-plastic" melt rheology. That is, their apparent viscosity decreases as the shear rate is increased. Clearly, with such materials it would be preferable to inject at high rates to take advantage of the lower viscosity that would assist in adequate filling of intricate mold cavities.

The requirements for adequate plasticization influence the machine settings for control of the injection phase. Here the two basic variables are injection rate and injection pressure. However, their effects must always be taken into account in conjunction with nozzle, gate, and cavity geometry. The effect of injection rate has already been considered in some detail. It should be added, however, that high rates through narrow nozzles and gates result in high shear rates, with consequent lowering of apparent viscosity, and high adiabatic heating (i.e., the charge becomes hotter as it is injected). This latter effect also contributes to a reduction in viscosity, but high shear rates and narrow flow passages also tend to increase fiber attrition. Thus, a working compromise must be established. The degree to which the various settings are exploited depends very much on the nature of the material to be molded. Thus, polyamides have a low melt viscosity and are relatively stable (in the absence of air and water) so that the charge may be fully plasticized in the barrel (i.e., rela-

tively high heat settings). Moreover, low back pressure, injection speed, and wide nozzle and gate minimize fiber breakdown. On the other hand, with a material such as polyacetal that is prone to degradation at elevated temperatures, it may be better to reduce barrel temperatures but arrange for some adiabatic heating on injection by using a higher injection rate and small nozzle and gate.

Injection pressure is of vital importance. Too little pressure will result in poor filling with consequent high shrinkage and poor surface finish, and, as a consequence, higher than optimum plasticization temperatures might have to be used. However, the effects of pressure are more subtle. At the pressures used in injection molding, 100–200 MPa (15,000–30,000 psi), most thermoplastics will be compressed to a significant degree, typically 1–5% by volume. This effect can be used to offset the inevitable shrinkage that occurs when the liquid polymer is cooled and solidifies. Thus, a liquid polymer freely solidifying at ambient pressure may shrink 5–10% by volume, but by filling the cavity at high pressures, more material is forced into the mold and, thus, shrinkage is reduced significantly. Shrinkage is also controlled by the feeding that occurs during the dwell phase.

Fibers and/or mineral fillers also serve to reduce the amount of shrinkage occurring when the molten polymer solidifies. This shrinkage reduction is generally greater than the amount that might be expected from a simple rule of mixtures calculation due to physical constraint effects. These effects are greater for fibers than for particulate fillers; they are a function of the fiber orientation and are not uniform.

The actual flow pattern of charge into the mold cavity determines the molecular and fiber orientation pattern in the molding, the shrinkage, and the level and distribution of residual stress in the part. It can be influenced by the speed of injection and by the effective injection pressure (i.e., the pressure in the mold cavity). In the simpler machines, the settings are achieved by simple flow restrictor and relief valves, but recently many machines have been fitted with feedback control systems. These systems use signals from pressure, temperature, and position transducers set at critical points in the runner or mold cavity so that injection rate and temperature can be controlled throughout the cycle. For instance, the initial fill rate may be low but then can be speeded up and the pressure increased gradually or stepwise according to a predetermined program. Such a

---

[6]It should be noted that special abrasion resistant materials are available for barrels (liners), screw and screw tips, and molds, which effectively minimize wear given the proper molding conditions. Use of such materials permits higher injection speeds.

system is highly desirable and affords the possibility of very precise control.

For molding fiber reinforced thermoplastics the following general rules should be observed:

a. Use temperature settings on the barrel and nozzle that ensure full plasticization of the charge before commencement of the injection stroke. Where possible, temperatures should be higher than for unfilled material.
b. Use slow screw rotation and low back pressure to reduce fiber attrition and machine wear.
c. Use slow injection rates and as high an injection pressure as practicable.
d. Allow adequate dwell time.
e. Follow appropriate procedures for drying and/or preheating the charge granules.

In practice, the adoption of these rules means that a machine can be used at only 50–70% of its rated capacity for unfilled polymers if high quality fiber reinforced parts are to be molded.

3.8.5 General Aspects of Mold Design

In this subsection we will be concerned with general principles of mold design as they affect the operation of the molding process and the quality of the parts molded. Fiber reinforced polymers have higher melt viscosity than unfilled materials so that higher pressures are needed to fill cavities of comparable intricacy. At high shear rates this effect is often minimized, but the effect can be exploited only at the expense of a higher degree of fiber attrition, as discussed in the previous subsection.

Ideally, molds should have generous gates and direct smooth-flowing runner systems. Where possible, multiple gating, with attendant weld line (knit line) problems, should be avoided. It is sometimes advantageous to use a single cavity mold on a smaller machine than to go for a multicavity design that calls for a more complicated gating and runner system. Multicavity molds, especially on larger machines, are, of course, used extensively to obtain high production rates.

When the design of the part is such that weld lines are inevitable, their effects can be minimized by ensuring that they occur in regions that are not critically stressed—by arranging that the material is still fully plasticized when contact occurs (preferably with some

induced turbulence in the critical region) and by setting a somewhat longer dwell time. This requirement means that the entire cavity must be filled and subjected to the full injection pressure *before* solidification commences. This technique often means that charge and mold temperatures must be raised, with inevitable extension of the total cycle time.

Shrinkage, which is lower in fiber filled materials, must be allowed for in determining draft angles and ejector pin positions. Molds must always be adequately vented. In general, undercuts are to be avoided.

In designing molds it is most important to ensure optimum cooling. The gate should not freeze too soon; otherwise feeding will be impaired. Apart from this problem it is, of course, desirable to cool as quickly as possible in order to minimize the cooling phase of the molding cycle, which inevitably determines the overall production rate.

Surface finish, another important consideration, depends on a number of factors, especially on the flow path of the charge in the mold, which is often complex and difficult to predict, and on the cavity wall temperature. The latter parameter must often be set considerably higher than that for unfilled polymers. This increased temperature has a drastic effect in increasing the cooling time; however, it is the penalty for increased surface finish.

Because mold filling and shrinkage are difficult to predict, it is always good practice to make undersized cavities in the first instance. These cavities may then be opened up when trial moldings have indicated the magnitude of shrinkage.

3.8.6 Special Considerations for Molding Filled Materials

The remarks in the previous subsections apply in some measure to all injection molding applications. We shall now consider the special character of filled thermoplastics, in general, and FRTP, in particular. Adding a mineral filler to a thermoplastic alters its properties. In general, the elastic modulus is increased, tensile strength is decreased, and shear and compression strengths are increased to a lesser degree, whereas ductility is decreased.[7] The improved stiffness and dimensional stability are generally the prime design motiva-

---

[7]Generally decreased elongation can adversely affect impact strength in certain fiber/resin systems.

tions for adopting such materials. From the processing point of view, the filled materials have lower heat input requirements and higher melt viscosity but exhibit much reduced shrinkage on molding and increased dimensional stability after molding—highly desirable characteristics for precision engineering usage.

In FRTP these improved properties are also influenced by the fiber distribution and orientation within the molding. If substantial preferred fiber orientation can be induced, properties such as stiffness and strength will be enhanced to a much greater degree in the principal orientation direction and, conversely, to a much smaller extent in directions normal to the orientation. At first sight it would seem that this effect could be exploited. However, the flow of FRTP (or indeed any plastic) into a mold during injection molding is a complex process, and it is very difficult to ensure that flow-induced orientation occurs in the optimum directions. In point of fact, the flow pattern does not differ much between filled and unfilled materials, but the extent of orientation-induced anisotropy is much greater in FRTP than in either unfilled or particulate filled materials. For this reason, particulate fillers may be preferred to fibers when the overriding consideration is the minimization of the undesirable effects of anisotropy, such as part warpage, which can be accentuated by flow orientation of fibers. In general, however, considerably superior properties can be achieved by fiber reinforcement with careful attention given to processing details that can minimize undesirable orientation effects. The anisotropy of the FRTP also extends to the "in-mold" and "post-molding" shrinkage. Shrinkage is generally much less in the "flow" direction than "across the flow." As in the case of unfilled materials, shrinkage is strongly influenced by the efficiency of mold filling and higher effective injection pressure. Longer dwell times generally reduce the overall level of shrinkage.

There will always be "frozen-in" residual stresses in an injection molded part; they can often be dissipated by a post-molding annealing operation. This stress relief is accompanied by further shrinkage, but thereafter the part will exhibit greatly improved dimensional stability.

When it is necessary to maximize precision and dimensional stability and to minimize warpage, it is necessary to maintain a positive control over all stages of the molding operation, from drying of the molding granules through the actual molding cycle and on to the final annealing operation.

3.8.7 Fiber Reinforced Thermoplastics (FRTP) Properties

FRTPs consist of short fibers dispersed in the thermoplastic matrix. The most important fiber used is *E*-glass, although high stiffness and strength graphite (carbon) fibers are assuming a more significant role. In the past, asbestos fibers were used for reinforcement of thermoplastics since they are relatively low cost and effective; however, their use is diminishing because of the health hazard associated with asbestos. Both *E*-glass and graphite fibers are produced in continuous tows or rovings and must be compounded with the thermoplastic to produce a moldable intermediate—usually molding granules. The properties of the molded part depend on the level of filling (i.e., weight or volume fraction of fiber), the fiber length distribution, the fiber orientation distribution, and the interfacial bond between the matrix and the fiber (see Volume 2). The theory of short-fiber reinforcement is dealt with in Volume 2, but it is pertinent to emphasize some critical considerations.

The maximum fiber content, V_f, that can be utilized in injection moldable systems is limited by the necessity for the compound to flow during the molding operation. There is no theoretical limit, but a V_f of 0.4 represents a usual upper limit; 0.25–0.35 is more typical when intricate components are to be molded. Likewise, the maximum fiber length *in the molded part* cannot usually exceed 10 mm (⅜ in); it is often only 0.1–0.3 mm (0.004–0.012 in). The fiber is reduced to these lengths either by deliberate chopping during the compounding operation or as a result of the shear and attrition induced during the compounding and molding operations. Long-fiber compounds with fiber lengths in the order of 12 mm have recently been considered by Crosby [2].

The effectiveness of the reinforcement is a function of V_f, the fiber length, and the orientation. For an array of equal length fibers all aligned in one direction, Young's modulus, E_c, of the composite is estimated by

$$E_c = E_f V_f [1 - (L_c/2L_f)] + E_m(1 - V_f) \qquad (3.8\text{-}1)$$

when $L_f > L_c$ or when $L_f < L_c$

$$E_c = \tau L_f V_f + E_m(1 - V_f) \qquad (3.8\text{-}2)$$

where

$$L_c = \frac{E_f \epsilon_{uf} r_f}{\tau} \qquad (3.8\text{-}3)$$

and E_f, E_m, E_c are the Young's moduli of fiber, matrix, and composite; L_f and r_f are the length and radius of the fiber, respectively; ϵ_{uf} is the ultimate tensile strain of the fiber; and τ is the shear strength of the fiber/matrix interface. When $L_f > L_c$, reinforcement is more efficient. A practical optimum is reached when $L_f \cong 5L_c$ and the consequent term $[1 - (L_c/2L_f)] \cong 0.9$, so that the reinforcement efficiency approaches 90% of that of a continuous fiber composite. In practical FRTP, L_c tends to be of the order of 100 μm so that optimum fiber lengths of 500 μm (0.02 in) would be appropriate. However, if the fibers are misaligned with respect to the loading direction, the efficiency decreases. This decrease may be conveniently dealt with by an orientation factor, C_o, which would be unity for the fully aligned case, 0.3 for the random in-plane case, and about 0.1 for the fully random distribution. Real composites also contain a distribution of fiber lengths so that either an average fiber length $\bar{L}_f$ or a summation over the whole fiber length distribution must be used (see Volume 2). In practice, the fiber length distribution can be determined quite conveniently, but the orientation distribution is more tedious to obtain.

Thus, we may arrive at an equation for composite stiffness of the form

$$E_c = C_o \left[\sum \frac{\tau L_i V_i}{2r_f} + \sum E_f V_j \left(1 - \frac{E_f r_f}{2L_j \tau} \right) \right]$$

$$+ E_M (1 - V_f) \qquad (3.8\text{-}4)$$

where the first term within the bracketed expression represents the contribution to stiffness of the fibers of length, $L_i < L_c$, and the second term that of the supercritical fibers, $L_j > L_c$. If $\bar{L}_f > L_c$, then a simpler expression is adequate:

$$E_c = C_o \left[E_f V_f \left(1 - \frac{E_f r_f}{2\bar{L}\tau} \right) \right] + E_m (1 - V_f)$$

$$(3.8\text{-}5)$$

This analysis indicates the importance of both fiber length and orientation on the stiffness of the composite. Short fibers (i.e., $L < L_c$) and even particulate fillers

can significantly improve the modulus of thermoplastics, but longer fibers are more efficient. Longer fibers also contribute more effectively to the enhancement of tensile strength and toughness.

Fiber orientation is clearly extremely important. It is influenced to some extent by V_f and L_f where high values of both parameters increase the degree of fiber alignment along flow contours. However, the flow field in molding is often very complex. In most practical moldings, the degree of alignment in the flow direction corresponds to values of C_o, in Equations (3.8-4) and (3.8-5), between 0.6 and 0.3. Therefore, if orientation can be exploited, there is a stiffness benefit of $\times 2$ over the random in-plane case ($C_o = 0.3$).

The critical length, L_c, is a function of the shear strength of the fiber/matrix interface, as can be seen from Equation (3.8-3). A closer examination shows that it is the fiber aspect ratio ($L_f/2r_f$) that is the critical factor, so that a smaller diameter fiber might be expected to be more effective. In practice, this effect is observed [3], but the extent of improvement is seldom sufficient to justify the use of especially fine fibers. Within the normal limits for commercial glass-fiber diameters, i.e., 7–20 μm (0.0003–0.0008 in), there is little difference in the properties of molded FRTP; therefore, in most instances, fibers in the range 10–13 μm (0.0004–0.0006 in) are used. Carbon fibers used for reinforcement are generally somewhat finer at 7–9 μm (0.0003–0.0004 in). The shear strength of the fiber/matrix interface depends on both chemical and mechanical factors. The mechanical effects arise from the shrinkage of the matrix onto the fiber when the thermoplastic cools from the melt. This shrinkage generates a radial pressure on the fiber from the surrounding matrix. The development of a mechanical bond is also influenced by the ability of the molten polymer to wet the fiber surface. Chemical interactions involve the formation of actual chemical bonds between the fiber surface and the polymer.

The indications are that strong mechanical bonds are formed between several polymers and graphite fibers. These bonds are particularly marked in the case of polyamides (nylons), acetals, polycarbonates, and similar strongly polar polymers. At this point, it should also be noted that the graphite fibers tend to be rougher than glass fibers. It is not clear whether actual chemical bonding occurs between these polymers and the graphite-fiber surface, but the net effect is that efficient reinforcement is achieved without recourse to complex surface pre-treatment of the fiber. All surface treatment that is usually done is to apply a standard mild surface oxidation treatment that removes loose debris and creates some active bonding sites on the fiber surface.

Glass fibers, on the other hand, generally require a treatment with a silane coupling agent if satisfactory interface properties are to be achieved [4]. This agent is incorporated in the size or finish that includes also a film-forming agent, which binds the fibers together into a strand. The combination of these two agents in the glass finish has important implications in the processing and properties of GFRTP. In general, a high strength interface can be achieved quite simply with the more polar resins, especially polyamides that are amenable to hydrogen bonding. On the other hand, non-polar materials such as polypropylene require elaborate modifications to both the polymer and the glass if satisfactory bonding is to be achieved. The so-called "chemically coupled polypropylene-glass compounds" are prepared from a polymer that incorporates a co-monomer additive to form bonding sites, and the glass is treated with a specially developed silane coupling agent.

3.8.8 Compounding FRTP

FRTP compounds may be prepared by a number of alternative routes, and the choice of route influences both the selection of the reinforcing fiber and the properties of the molded parts. The main compounding routes are as follows:

a. Extrusion compounding and pelletizing (chopped fiber)
b. Extrusion compounding and pelletizing (continuous fiber)
c. Extrusion coating (or melt coating) and chopping (continuous fiber)
d. Direct molding of dry blended chopped fiber and polymer fiber

In processes "a" and "d", milled fiber may be substituted for chopped fiber. Milled fibers are much shorter than the normal "chopped" product. The processes are compared in the flow sheet presentation of Figure 3.8-7.

The most widely used process is "a". Fiber rovings are chopped into suitable lengths, usually 6 mm (¼ in), and dry blended with the polymer, which is usually in the form of standard molding granules of 2–4 mm

(0.08–0.16 in) diameter. The blend is charged into the hopper of a single or twin screw extruder and extruded through a "spaghetti" die and pelletized. The screw plasticization results in a thorough mixing of glass fibers and polymer so that the molding pellets produced are homogeneous. The fiber strand integrity has an important influence on the quality of the molding pellets produced. During extrusion, the feed undergoes high shear stresses that tend to break up the fiber strands into individual filaments and subsequently break these filaments into even shorter lengths. By using a strong film former in the fiber size, the chopped rovings are bound into clumps of fibers that resist the shearing action of the extruder to some extent. Sometimes clumps maintain their integrity through the compounding process, and full fiber dispersion is achieved only during the final injection molding operation. Clearly, a balance has to be achieved. Clumped fibers in the final molding detract from its appearance and can lead to unsatisfactory performance, which would be the effect of too strong a film former or insufficient shear during compounding and molding. The other extreme results in a very high level of fiber breakages, very short fibers [$L_f < 0.1$ mm (0.004 in)], and inferior mechanical properties. A high degree of strand integrity is essential at the dry blending stage of the operation. Otherwise, it is difficult to feed the extruder adequately due to "fuzz-balling" of the fibers—a special hazard when blending high V_f mixtures. Extrusion compounding may be carried out using a variety of extruder types. The twin screw machines are generally preferred because they have greater production capacity and the degree of shear can be more easily controlled.

Method "b" is a development of "a" and involves the use of a two-stage vented extruder (either single or double screw although the latter one would, again, be preferred). The fibers are introduced directly into the melt at the decompression section. Either chopped fiber rovings or continuous rovings may be fed in this way. In the latter case, the rovings are "chopped" solely by the action of the extruder. The advantage of this method is that feeding problems associated with the dry chopped fiber/polymer blends are eliminated and screw and barrel wear are reduced. Furthermore, less fiber attrition occurs and, hence, longer fibers may be retained in the molding granules.

In the extrusion coating (method "c"), continuous rovings of fiber are passed through a cross-head die on an extruder. The fibers are coated with polymer in a

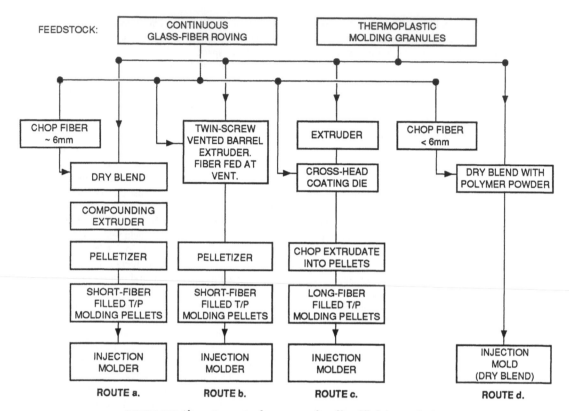

FEEDSTOCK:

FIGURE 3.8-7. Alternative routes for compounding fiber filled thermoplastics.

similar manner to a wire insulating process. The coaxial product is then chopped into molding pellets. The advantage of this method is that the fibers suffer no attrition and the fiber length is the same as the pellet length [typically 4–10 mm (⅛–½ in)]. The process, however, does not fully impregnate the fiber strands so that full dispersion of the fibers in the matrix must be accomplished during the final molding operation, which requires careful attention to machine settings and gate geometry. An advantage of this method is that long-fiber-containing moldings can be made. Typically, fibers of up to 4–6 mm (⅛–¼ in) survive through the final molding operation, whereas, in other processes ("a" and "b"), the fibers have already been reduced to lengths of less than 1 mm (0.04 in) in the initial compounding operation. Materials manufactured by this route are available commercially, but they comprise only a small proportion of the total market. They are more difficult to mold successfully than are conventional compounds. One of the main problems is warpage induced by the high degree of anisotropy.

Another variant of this process is to coat fiber rovings by drawing them through a bath of molten polymer (or through a solution of polymer). This is a very limited option, as few thermoplastics remain chemically stable for sufficient periods at the temperatures required to achieve a sufficiently low viscosity for the coating operation.

The final option, "d", is to dispense with the separate compounding operation and to charge a dry blend of chopped roving and polymer directly into the injection molding machine. The compounding operation occurs during the normal plasticization cycle. This practice is difficult to control, especially when a variety of molding machines is used. However, it has been used with some success for glass-coupled/polypropylene blends utilizing fairly low fiber volume fractions ($V_f \cong 0.25$) and equipment designed especially for the process.

3.8.9 Mold Filling and Flow-Induced Fiber Orientation

In laminar flow both fibers and polymer molecules tend to align with the streamlines. Polymer melts have high viscosity so that the laminar to turbulent flow transition would be expected to occur at relatively high shear rates. This occurrence is generally true, but the flow of a thermoplastic into a mold is a complex process; consequently, the flow-induced orientation in the molding is difficult to predict.

The sprue and runner system leading from the injection nozzle constitutes a parallel or convergent duct leading to the mold gate, which normally has the least cross section of the whole flow path. Thereafter, flow is essentially divergent. Convergent and parallel flow induces orientation parallel to the streamlines, but divergent flow tends to introduce fiber orientation normal to the flow direction.

Injection mold cavities are usually in the form of plate-like sections with stiffening ribs. Good design practice maintains section thickness as constant as possible, within the constraints imposed by the design, to ensure uniform cooling rates throughout the molding.

Thus, the most common flow pattern is from a converging or parallel runner channel through a narrow gate, which then diverges into a thin plate form and gives a two-dimensional divergent flow path.

An idealized flow would be as shown in Figure 3.8-8, where the mold fills uniformly from the gate and the regions remote from the gate are last to fill. This ideal flow is seldom achieved due to the high rate of flow of material through the gate, which might result in the flow pattern depicted in Figure 3.8-9(a). This behavior is termed "jetting." In extreme cases the material will "jet" in a thin stream right across the mold, which is then filled by folding of the jet [Figure 3.8-9(b)]. This behavior is more likely if high injection rates are used with highly thixotropic fluids such as highly filled thermoplastics.

The flow pattern is influenced by friction and chilling of the charge at the mold surfaces. This effect is most pronounced in the orientation distribution through the thickness of a plate-type molding. In the middle of the section the flow is relatively unrestricted, but at the surfaces the charge is chilled and sticks to the mold. These conditions result in a fan-like flow through the thickness direction which is superimposed on the in-plane flow pattern. Under uniform flow, e.g., Figure 3.8-10, these conditions can result in either a three- or a five-

zone fiber orientation distribution through the section thickness. The fibers in the center section are oriented transverse to the principal flow direction, whereas the surface and subsurface fibers are aligned with the flow direction [5].

Another phenomenon is that a thin layer of melt may solidify at the mold surface and then be displaced along the surface by the flowing charge. This layer gives rise to surface imperfections and orientation discontinuities. Irregular flow patterns can also result in unpredicted weld lines where streams impinge.

Within the restrictions imposed by the design function of the molding, there are ways of establishing a measure of control over the orientation distribution in the molding:

a. The detail design of the cavity
b. The position and form of runners and gate(s)
c. Injection rate
d. Temperature of the charge
e. Mold temperature

These measures of control are discussed in the following segments.

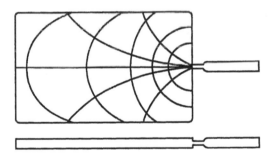

a. IDEAL FILLING FROM SINGLE END GATE

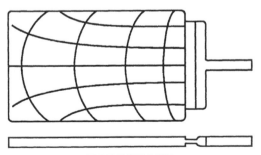

b. IDEAL FILLING FROM FLASH GATE

FIGURE 3.8-8. Flow contours and orientation directions.

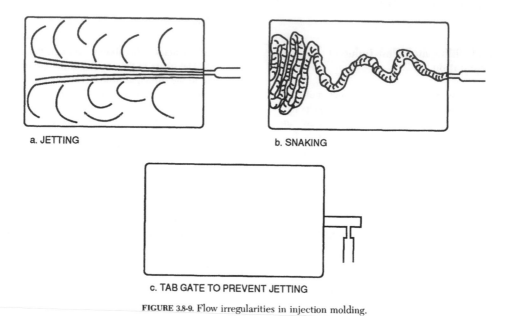

a. JETTING

b. SNAKING

c. TAB GATE TO PREVENT JETTING

FIGURE 3.8-9. Flow irregularities in injection molding.

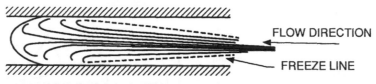

MOLD WALLS

FLOW DIRECTION

FREEZE LINE

a. TYPICAL FLOW PATTERN THROUGH THICKNESS OF MOLDING

SURFACE

1
2
3
2
1

1. AXIAL
2. RANDOM IN-PLANE
3. TRANSVERSE

SURFACE

b. 5-LAYER ORIENTATION DISTRIBUTION

SURFACE

1
3
1

1. AXIAL
3. TRANSVERSE

SURFACE

c. 3-LAYER ORIENTATION DISTRIBUTION

FIGURE 3.8-10. Through thickness flow patterns and fiber orientation distributions.

Cavity Design

Wherever possible, the part should be designed to have uniform wall thickness, and transitions from thin to thick sections in critical areas should be avoided. Careful attention should be given to the positioning of holes and cut-outs that might result in undesirable weld lines [6].

Runners and Gates

Runners should be of "full-round" form to minimize heat losses and should be designed to avoid sharp changes of direction. They should be of generous cross section and should be kept as short as feasible. Generally, multiple gating should be avoided, as weld lines are then inevitable. In some instances, however, weld line position can be controlled more easily in complex moldings by multiple gating.

In general for fiber filled thermoplastics, gate dimensions should be generous, i.e., $>50\%$ of section thickness. The gate should be positioned to encourage flow along the principal stress directions, so that flow-induced anisotropy can be exploited. For wide sections, flash gates are often preferred.

Injection Rate

Injection rate has considerable influence over the flow pattern. Fast rates encourage jetting and turbulence, which are always undesirable. On the other hand, if the rate is too slow, it is difficult to fill complex cavities. For this reason, interactive injection rate control is invaluable for molding filled thermoplastics. The initial rate is kept low to prevent jetting, but the rate is increased as the stroke progresses and then reduced again as the cavity fills. Pressure transducers located at critical points in the cavity and/or runners and displacement transducers on the ram provide the necessary input data for control. The great advantage of this system over traditional passive controls is that, once a satisfactory molding cycle has been established, it can be consistently maintained throughout the molding cycles.

Charge Temperature

Charge temperature is important in that it controls the viscosity of the charge during molding filling. In this sense, charge temperature and injection rate must be controlled together. The other factor influenced by charge temperature is the rate of cooling and solidification of the polymer in contact with the mold surface. Higher temperatures permit lower injection rates and avoid the problems due to local solidification before the cavity is filled.

Mold Temperature

Mold temperature controls the effects due to surface heat transfer. A higher temperature allows for lower injection rates, which often result in more controlled filling of the cavity. The penalty is a longer molding cycle. High mold temperatures usually result in better surface finish and a less pronounced surface layer.

3.8.10 Precision Molding

One of the major attractions of thermoplastics is the ability to mold complex parts at high rates of production. This advantage can be seriously eroded, however, if control of dimensions is not consistent. The main reason for lack of dimensional stability is the shrinkage which occurs when a polymer solidifies in the mold. There is also the possibility of post-molding shrinkage when the part is exposed to ambient or elevated temperatures over an extended period of time. Post-molding shrinkage is the consequence of the relaxation of frozen-in stresses and molecular orientation and, in the case of semicrystalline polymers, further crystallization. Many polymers also absorb significant quantities of water from the atmosphere, which may cause swelling and also influence relaxation processes and crystallization by plasticization.

In general, unfilled polymers shrink by about 5% on solidification, but filling with mineral or fibrous fillers reduces this shrinkage to 1% or less in some cases. However, with both filled and unfilled materials, shrinkage can be reduced by optimization of the molding conditions. Long-term dimensional stability may often be further enhanced by a post-molding annealing heat treatment.

Whereas recommendations will vary considerably for different molding materials, the following general principles apply.

To reduce in-mold shrinkage:

a. Pre-dry molding granules (preheating is also sometimes recommended).

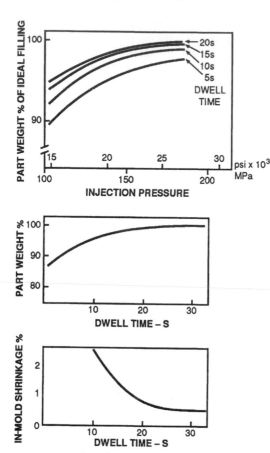

FIGURE 3.8-11. Effect of injection pressure and dwell time on mold filling and shrinkage for typical FRTP.

b. Use high injection pressure.
c. Use high mold temperature.
d. Use recommended melt temperature.
e. Use adequate dwell (screw forward) time.

To further improve dimensional stability:

f. Anneal in oil bath or air circulating oven.

In order to establish optimum molding conditions, simple trials should be run. Melt temperature should be maintained at the material supplier's recommended value. If possible, this temperature should be measured by a thermocouple inserted into the melt stream just ahead of the nozzle. Then injection pressure and dwell

time should be gradually increased until the weight of the molding has stabilized at its highest level (see Figure 3.8-11). Mold temperature should normally be set towards the top of the recommended range for the material.

It should be noted that shrinkage will be greater in thicker sections. In unfilled thermoplastics, shrinkage is greater in the direction of flow, but in fiber filled materials the reverse is true, i.e., greater shrinkage occurs normal to the fiber orientation direction.

With appropriate machine settings, it is possible to mold fiber reinforced thermoplastics and achieve in-mold shrinkages of less than 0.5% and tolerances of ±0.1% without post-mold annealing. Precision parts subjected to elevated in-service temperatures (i.e., above 50°C) will, however, usually benefit from an annealing heat treatment.

Where the highest levels of dimensional stability in the finished part are required, base thermoplastics with low water absorption should be selected (e.g., polyamide 11 rather than polyamide 6.6).

Note that graphite fibers have a negative coefficient of thermal expansion. This property can be exploited to produce moldings with almost zero expansion over the ambient temperature range in the direction of fiber orientation.

3.8.11 Fiber Reinforced Thermoplastics (FRTP) Summary

Virtually all thermoplastics materials can and have been reinforced by the addition of short fibrous fillers. The technologically important fillers at the present time are glass and graphite fibers.

The general benefits of fiber reinforcement may be summarized as follows:

a. Increased stiffness
b. Increased strength
c. Decreased in-mold shrinkage
d. Decreased coefficient of thermal expansion
e. Increased dimensional stability
f. Enhanced elevated temperature performance

These benefits are usually at the expense of a reduction in ductility and often of toughness. These effects differ quite markedly in different thermoplastics systems.

Soft and ductile polymers such as polyethylene and polypropylene show reduced toughness. Semicrystal-

Table 3.8-1a. Properties of fiber reinforced thermoplastics.

1A Polymer	2A Glass Fiber Content W%	V%	3 Specific Gravity	4 Typical Melt Temperature (°C)	5 Typical Mold Temperature (°C)	6 Typical Mold Shrinkage (%)	7 Coefficient of Thermal Expansion (10^{-5}/°C)	8 HDT (°C)
1. Polyethylene (HD)	20	9	1.10	230	40	0.45	5.4	125
2. Polyethylene (HD)	40	20	1.28	230	40	0.35	4.3	125
3. Polypropylene	20	8	1.04	245	40	0.45	4.3	140
4. Polypropylene (Chemically Coupled)	20	8	1.04	245	40	0.45	4.5	150
5. Polypropylene (Chemically Coupled)	40	19	1.22	245	40	0.35	2.7	155
6. Nylon 6	40	23	1.46	280	95	0.40	2.2	215
7. Nylon 6.6	20	10	1.28	295	105	0.60	4.1	252
8. Nylon 6.6	40	23	1.46	295	105	0.50	2.5	260
9. Nylon 6.10	40	22	1.41	280	95	0.40	2.2	215
10. Nylon 11	30	15	1.26	250	90	0.50	5.4	173
11. Acetal Homopolymer	20	12	1.56	200	95	0.5	3.6	157
12. Acetal Copolymer	30	19	1.63	210	95	0.5	4.3	162
13. Acetal (Chemically Coupled)	30	19	1.63	210	95	0.3	4.0	165
14. Polystyrene	40	22	1.38	245	65	0.10	3.4	104
15. SAN	40	22	1.40	260	90	0.10	2.7	104
16. ABS	40	22	1.38	260	90	0.15	2.2	107
17. Modified PPO	40	22	1.38	325	105	0.20	1.8	157
18. PETP	30	18	1.61	302	95	0.40	2.7	225
19. PBTP	40	26	1.62	240	95	0.35	1.9	232
20. Polysulfone	40	26	1.55	360	150	0.20	2.2	174
21. Polyethersulfone	40	26	1.68	360	150	0.15	2.9	215
22. PPS	40	26	1.65	320	120	0.25	2.7	263
23. Polycarbonate	20	10	1.34	315	120	0.25	2.7	150
24. Polycarbonate	30	17	1.43	315	120	0.20	2.3	150
25. Polycarbonate	40	24	1.52	3.5	120	0.20	1.8	150
Graphite (Carbon) Fiber Filled Materials								
	Graphite Fiber Content							
26. Nylon 6.6	30	21	1.28	295	105	0.25	1.9	257
27. PBTP	30	24	1.47	240	95	0.30	0.9	220
28. Polysulfone	30	24	1.37	360	150	0.15	1.08	185
29. PPS	30	24	1.45	320	120	0.15	1.08	263

1. Fiber Content: Most suppliers quote weight %. In Column 2 volume % is also given. This has been calculated from the quoted density of the unfilled polymer with the density for E-glass being taken as 2.54 g/cm^2 and for graphite fiber (type A) as 1.8 g/cm^3.
2. Melt Temperature and Mold Temperature: Columns 4 and 5 are typical values taken from suppliers' recommendations.
3. Mold Shrinkage: This is the shrinkage for a 6 mm thick section in direction of flow. ASTM Method D955 was used where available; otherwise suppliers' data were used.
4. Coefficient of Thermal Expansion: ASTM Method D696 in flow direction is applied.
5. HDT – Heat Distortion Temperature: ASTM Method D648 at 1.81 MPa (264 psi).
6. Water Absorption: This is equilibrium value for immersion at 23°C (ASTM D570). Values from different suppliers varied widely.
7. Flexural Modulus: Measured at 20–23°C (ASTM D790).
8. Tensile Strength: Measured at 20–23°C (ASTM D638).
9. Tensile Elongation: Measured at 20–23°C (ASTM D638). Data from different suppliers vary considerably.
10. Notched Izod Impact Strength: Most values quoted were of tests according to ASTM D256 on notched 6 mm (¼ in) bars.

Table 3.8-1b. Properties of fiber reinforced thermoplastics.

1B Polymer	2B Glass Fiber Content W%	V%	9 Water Absorption (Max) %	10 Flexural Modulus (GPa)	11 Tensile Strength (MPa)	12 Tensile Elongation (%)	13 Notched Izod Impact (J/m)
1. Polyethylene (HD)	20	9	0.1	4.0	55	2.5	50
2. Polyethylene (HD)	40	20	0.3	7.5	80	2.5	70
3. Polypropylene	20	8	0.02	4.0	63	2.5	75
4. Polypropylene (Chemically Coupled)	20	8	0.02	4.0	79	4	90
5. Polypropylene (Chemically Coupled)	40	19	0.09	7.0	103	4	100
6. Nylon 6	40	23	4.6	10.5	180	2.5	150
7. Nylon 6.6	20	10	5.6	9.0	130	3.5	100
8. Nylon 6.6	40	23	3.0	15	210	2.5	136
9. Nylon 6.10	40	22	1.8	9.0	210	2.5	170
10. Nylon 11	30	15	0.4	3.2	95	5	–
11. Acetal Homopolymer	20	12	1.0	4.3	60	7	40
12. Acetal Copolymer	30	19	1.8	9.0	90	2	40
13. Acetal (Chemically Coupled)	30	19	0.9	9.7	135	4	95
14. Polystyrene	40	22	0.1	11.3	103	2.5	60
15. SAN	40	22	0.28	13.4	128	2.5	60
16. ABS	40	22	0.5	7.6	110	3.5	70
17. Modified PPO	40	22	0.09	8.6	135	3.5	80
18. PETP	30	18	0.24	8.3	130	4	85
19. PBTP	40	26	0.4	9.6	150	4	155
20. Polysulfone	40	26	0.6	11.0	138	2	100
21. Polyethersulfone	40	26	–	11.0	205	–	80
22. PPS	40	26	0.06	12.5	160	3	80
23. Polycarbonate	20	10	0.19	5.8	110	6	180
24. Polycarbonate	30	17	0.18	8.2	127	5	190
25. Polycarbonate	40	24	0.16	10.3	145	4	200
Graphite (Carbon) Fiber Filled Materials							
	Graphite Fiber Content						
26. Nylon 6.6	30	21	2.4	20.0	240	3.5	75
27. PETP	30	24	0.3	13.8	138	2.5	60
28. Polysulfone	30	24	0.4	14.0	148	2.5	60
29. PPS	30	24	0.1	16.9	186	2.5	55

1. Fiber Content: Most suppliers quote weight %. In Column 2 volume % is also given. This has been calculated from the quoted density of the unfilled polymer with the density for *E*-glass being taken as 2.54 g/cm² and for graphite fiber (type A) as 1.8 g/cm³.
2. Melt Temperature and Mold Temperature: Columns 4 and 5 are typical values taken from suppliers' recommendations.
3. Mold Shrinkage: This is the shrinkage for a 6 mm thick section in direction of flow. ASTM Method D955 was used where available; otherwise suppliers' data were used.
4. Coefficient of Thermal Expansion: ASTM Method D696 in flow direction is applied.
5. HDT – Heat Distortion Temperature: ASTM Method D648 at 1.81 MPa (264 psi).
6. Water Absorption: This is equilibrium value for immersion at 23°C (ASTM D570). Values from different suppliers varied widely.
7. Flexural Modulus: Measured at 20–23°C (ASTM D790).
8. Tensile Strength: Measured at 20–23°C (ASTM D638).
9. Tensile Elongation: Measured at 20–23°C (ASTM D638). Data from different suppliers vary considerably.
10. Notched Izod Impact Strength: Most values quoted were of tests according to ASTM D256 on notched 6 mm (¼ in) bars.

line engineering thermoplastics (e.g., polyamides and polyacetals) are severely embrittled by fiber reinforcement (as measured by elongation), although notch sensitivity is often less marked. The toughness of these materials is often enhanced if longer fibers are incorporated. The brittle, glassy materials (e.g., polystyrene, polymethylmethacrylate) are both stiffened and toughened by fiber additions.

In Tables 3.8-la and 3.8-lb, typical properties of FRTPs are given. These have been extracted from manufacturers' brochures and the open literature and have been rounded off or averaged for this presentation. It should be noted that the data are based on test bar results and that the properties of moldings can (and do) differ quite markedly from these values, which should be considered only a guide and not design values.

3.8.12 References

1. ENGLISH, L. K. "Fabricating the Future with Composite Materials, Part I: The Basics," *Materials Engineering*, 4:15 (September 1987).

2. CROSBY, J. M. "Long Fiber Molding Materials," in *Thermoplastic Composite Materials*, L. A. Carlsson and R. B. Pipes, eds., Elsevier (1989).

3. SATO, N., T. Kurauchi, S. Sato and D. Kamigaito. "Reinforcing Mechanism by Small Diameter Fiber in Short Fiber Composites," *J. Composite Materials*, 22:850 (1988).

4. HALL, D. *An Introduction to Composite Materials.* Cambridge University Press (1985).

5. GILLESPIE, J. W., J. A. Vandershuren and R. B. Pipes. "Process Induced Fiber Orientation: Numerical Simulation with Experimental Verification," *Polymer Composites*, 6(2):82 (1985).

6. GASTROW, H. "Injection Molds—102 Proven Designs," K. Stoeckhert, ed., Oxford University Press (1983).

3.8.13 Bibliography

Theory of Discontinuous Fiber Composites

BOWYER, W. H. and M. G. Bader. "On the Reinforcement of Thermoplastics by Imperfectly Aligned Discontinuous Fibres," *J. Mater. Sci.*, 7:1315 (1972).

CHEN, P. E. "Strength Properties of Discontinuous Fibre Composites," *Pol. Eng. Sci.*, 11:51 (1971).

COX, H. L. "The General Principles Governing the Stress Analysis of Composites," *Fibre Reinforced Materials—Design and Engineering Applications*, Inst. Civ. Eng., London (1977).

CURTIS, P. T., M. G. Bader and J. E. Bailey. "The Stiffness and Strength of a Polyamide Thermoplastic Reinforced with Glass and Carbon Fibres," *J. Mater. Sci.*, 13:377 (1978).

HALPIN, J. C. and N. J. Pagano. "A Laminate Approximation for Randomly Orientated Fibrous Composites," *J. Composite Materials*, 3:720 (1969).

HALPIN, J. C. and J. L. Kardos. "The Halpin-Tsai Equations— A Review," *Pol. Eng. Sci.*, 16:355 (1970).

HALPIN, J. C. and J. L. Kardos. "Strength of Discontinuous Reinforced Composites: 1. Fiber Reinforced Composites," *Pol. Eng. Sci.*, 18:459 (1975).

LEES, J. K. "A Study of the Tensile Strength of Short Fiber Reinforced Plastics," *Pol. Eng. Sci.*, 8:195 (1968).

Properties of Fiber Reinforced Thermoplastics

BADER, M. G. and W. H. Bowyer. "The Mechanical Properties of Thermoplastics Strengthened by Short Discontinuous Fibres," *J. Phys: D*, 5:2215 (1972).

DARLINGTON, M. W. and P. L. McGinley. "Fibre Orientation Distribution in Short-Fibre Reinforced Plastics," *J. Mater. Sci.*, 10:906 (1975).

Fiber/Matrix Interface

BROUTMAN, L. J. "Mechanical Requirements of the Fiber-Matrix Interface," SPI 25th Ann. Tech. Conf. RP/CD, Section 13-B.

PLUEDDEMAN, E. P. "Adhesion through Silane Coupling Agents," SPI 25th Ann. Tech. Conf. RP/CD, Section 13-D.

YAMAKI, J.-I. "Tensile Strength of Discontinuous Fibre Composites—the Effect of Fibre-Matrix Interface Strength," *J. Phys: D*, 9:115 (1976).

Processing

BADER, M. G. and W. H. Bowyer. "An Improved Method of Production for High Strength Fibre-Reinforced Thermoplastics," *Composites*, 150 (July 1973).

CLOUD, P. J. and F. McDowell. "Reinforced Thermoplastics: Understanding Weld Line Integrity," *Plast. Technology* (August 1976).

CLOUD, P. J. and R. E. Schulz. "A Primer on Reinforcing and Filling I. M. Plastics," *Plastics World* (September 1975).

MENGES, G. and J. Lutterbeck. "New Developments and Trends in the Field of Extrusion and Injection Moulding," *Plast. and Rubber Int.*, 4(2):59 (1979).

STADE, K. H. "New Achievements in Compounding Glass Fiber-Reinforced Thermoplastics," *Reinforced Thermoplastics Symposium*, Plastics Institute, London (1975).

WOOD, R. "Continuous Compounding Equipment: Single and Twin Screw Extruders," *Plast. and Rubber Int.*, 4(5):207 (1979).

SECTION 3.9

Injection Molding Fiber Reinforced (F.R.) Thermosets and BMC

3.9 INJECTION MOLDING FIBER REINFORCED (F.R.) THERMOSETS AND BMC

M. G. BADER

3.9.1 Introduction

This section should be read in conjunction with sections 3.6 and 3.8 (Compression and Transfer Molding of Reinforced Thermoset Materials, and Injection Molding).

Injection molding was originally developed for thermoplastics using simple plunger type machines. Subsequently, machines based on the reciprocating screw plasticizer principle were developed. Until comparatively recently, it was considered that thermosets were not suitable for injection molding because of the risk of the material curing in the screw or plunger mechanism. The prospect of cleaning the screw, barrel, and nozzle filled with a fully cured thermoset is daunting, and the avoidance of this problem is central to the development of the technology of injection molding thermosets.

The state of the art has progressed to such an extent that injection molding is now the most important and fastest growing method for molding conventional thermosets (English, 1987).

3.9.2 Fundamental Principles

Whereas thermoplastics are melted and injected into a relatively cold mold where they are allowed to solidify, thermosets are injected into a hot mold where they cure and are ejected while still hot. Clearly, the material must be plasticized before it may be injected, and plasticization must be achieved without premature cross-linking or curing reactions. Thus, it is essential that the compound be plasticized to a sufficient extent at a temperature below that at which rapid cure occurs, and that cure not intervene over realistic residence times in the barrel of the injection molding machine.

Once the charge has been injected into the hot mold, it is desirable that it cure and be ejected in as short a time as possible, in order to achieve the maximum production rate. It should be noted that the cycle time is not dominated by the "cool in-mold" phase, as in the case of thermoplastic materials, but rather by the cure time.

3.9.3 Thermosetting Compounds for Injection Molding

Ideally, the base polymer should soften at a low temperature to a low viscosity liquid and be capable of being held in this condition for as long as possible without significant cure (i.e., without gelling). On raising the temperature to the cure temperature, it should then fully cure as rapidly as possible with the minimum evolution of volatiles. These ideals are most nearly approached when there is a large temperature differential between the plasticization and cure temperatures. Low melt viscosity may be achieved by using resins with a lower degree of advancement (see subsection 3.6.2); however, this generally means that a relatively larger quantity of volatile condensate will be evolved during the final cure. Likewise, a lower total filler content will give a lower viscosity melt but at the expense of higher volatiles and higher shrinkage during cure. Thus, it is necessary to balance the flow requirements of the injection process and the volatiles and shrinkage.

Typically, special grades of most thermosetting polymers that are produced for injection molding are similar to the high flow end of the ranges produced for compression and transfer molding. Most of the major thermoset materials are produced in injection moldable grades, which include the phenolics, amines, and alkyds with the so-called DAP compounds perhaps being among the most amenable to the process.

As in many thermoset systems, a large proportion of filler is included in the compound. Conventionally, particulate fillers are used, but for high-performance compounds a proportion of chopped or milled glass fiber may be included. These fibers are very short, typically less than 2 mm (0.08 in). Even if longer fibers are added into the compound, they are broken up during the plasticization and molding operations. Fibrous fillers increase the melt viscosity of the compound to a greater extent than particulate fillers; hence, fiber reinforced compounds are often more difficult to process than simple particulate filled materials.

3.9.4 Injection Molding Machines for Thermosets

The overall operating principle of the machine is similar to that described in section 3.8 for thermoplastics. However, the screw is generally shorter (lower L/D) and features a lower compression ratio. The screw and its associated rotating gear need to be more robust than those used for thermoplastics, owing to the higher melt viscosities of filled thermosets. The barrel is operated at a lower temperature, typically 70–100°C (160–212°F), for conventional thermosets; therefore, water or oil jackets often replace the electric resistance band heaters, which are common on thermoplastics processing equipment. The use of a liquid jacket ensures more even heating and eliminates the possibility of hot spots that might trigger premature cure.

Nozzles are generally of the open type, without positive shut-off valves, and the machines operate against low back pressures. The mold is heated by steam, recirculating oil, or electrical heaters to the cure temperature—typically 160–180°C (320–360°F). Runners may be either "hot" or "warm." In the former case, the material in the runner cures to a solid and must be cleared and scrapped before the next operation. (A small proportion of regrind can sometimes be tolerated.) In the "warm" runner system the material is held below the cure temperature in the plastic condition and is injected with the next charge. There is no scrap from this source. However, the warm runner system is more critical to operate and is usually employed only for "easy to mold" operations.

3.9.5 The Injection Molding Operation

The charge is normally predried, and sometimes preheated, and then is placed in the feed hopper. Conventional thermoset materials are supplied as granules, which flow into the throat of the screw without difficulty. Some compounds, however, may need to be force

charged, usually by a screw charging mechanism set up in the hopper.

The charge is picked up by the screw and transported through the barrel, where it is heated and plasticized by a combination of external heating from the barrel and mechanical work by the screw. As is the case when molding thermoplastics, the plasticized material is transported to the front end of the barrel and the screw retracts under the moderate back pressure to accommodate it.

At the end of the plasticization stage, the screw rotation ceases and the screw is driven forward to inject the charge into the mold. The injection pressure is generally much higher than that used for compression or transfer molding but lower than that used in molding thermoplastics—typically in the range 50–100 MPa (7.5–15 ksi). Injection rate control is critical because, at high shear rates, the temperature of the charge may be raised by a considerable amount, and this effect may be exploited to speed up the process of raising the charge temperature from the plasticizing range to the cure temperature—typically, a rise of some 70°C (140°F). This heating is more marked when a high injection speed and a small gate are combined. However, this combination also maximizes fiber attrition when fiber reinforced grades are molded, so this technique for speeding up the molding cycle must be used with caution. Also, if the charge is overheated, burning (degradation) and excessive evolution of volatiles may result.

The melt rheology of thermoset compounds is complex and not adequately understood. The compounds are highly filled and tend to exhibit extreme pseudoplastic behavior (i.e., apparent viscosity decreases as shear rate is increased). When subjected to high rates of flow through narrow passages (i.e., runners and gates), the material adjacent to the walls is subjected to a very high shear rate that gives rise to a "plug flow" behavior. When this behavior happens, the charge "snakes" into the mold cavity (see Figures 3.8-8 and 3.8-9) and results in a very irregular filling pattern. It is impossible to control the fiber orientation distribution in moldings when this plug flow occurs, and very variable properties result. It is, therefore, often desirable when molding fiber filled compounds to sacrifice the ultimate cycle rate in order to get a more uniform and predictable filling pattern. This pattern is particularly important in the case of fiber filled materials owing to the fiber orientation induced mechanical anisotropy characteristic of these compounds.

The final stage of the molding cycle is a cure that takes from a few seconds upwards. Fast curing compounds are available; additional fast cure is aided by use of high mold temperatures and fast injection rates. Thin wall sections cure faster than thick ones, and it is easier to vent such moldings. If temperatures are too high, blistering and burning can occur; furthermore, if parts are ejected too soon, they may be nonuniformly cured.

Adequate venting of volatiles is a requirement of the injection molding process. In part, venting is accomplished via venting passages cut into the mold, ejector pins, and core pins (as discussed in the following subsection). These provisions alone are sometimes inadequate to deal with the volume of gas produced, and in these cases, the mold must be opened once or twice during the cure period. Ideally, the mold pressure may be relaxed as soon as the charge has gelled. (Clearly, the injection pressure must be relaxed before the mold is opened at all.) A machine fitted with comprehensive interactive controls is preferred, and venting may then be controlled by monitoring the cavity pressure.

3.9.6 Mold and Cavity Design

Because of the high temperatures and pressures used in the molding cycle and the abrasive nature of filled thermoset compounds, it is desirable that molds be constructed from steel and that the faces of the cavities, gates, and runners be hardened. Highly polished working surfaces reduce friction and wear, as well as improve the surface finish of the product. For long production campaigns, hard surfacing processes should be considered (e.g., hard chromium plating, nitriding, or spray- or weld-deposited hard metals) for both cavity and runner areas. It can be advantageous to design the mold with a view to easy replacement of the gate where maximum wear would be encountered.

Apart from these considerations, the mold and cavity design requirements are very similar to those for injection molding thermoplastics. Ideally, moldings should have adequate draft (2° minimum) to ensure easy demolding and uniform section thickness so that filling and cure rates are uniform. No undercuts can be tolerated because cured thermosets tend to be brittle. Small components may be molded in multicavity molds, but when hot runners are used, this generally results in a greater material loss in scrapped runners and sprues.

For thermoset molding, adequate venting is essential. Venting passages may be cut into the parting line and into ejector and core pins. Failure to vent sufficiently may result in blistering or complete disintegration of the molding as the trapped gases expand when the mold is opened.

Hot runners should be of "full round" form and be designed for minimum restriction to flow. Gates are generally of larger dimensions than those for thermoplastics, particularly for fiber filled grades. Otherwise, excessive fiber breakage will occur.

Shrinkage in the mold is less of a problem for thermosets than it is for thermoplastics, since the high proportions of fillers used reduce the thermal contraction considerably.

Metal and/or ceramic inserts may conveniently be molded into thermosets. When these inserts are used, it is necessary to consider the effects of differential shrinkage strains between the insert and the molding. Extra material should be provided in the critical regions, particularly when the insert is to act as a point for introduction of the load into the molded part in service (e.g., a threaded screw bush or a bearing insert).

3.9.7 Summary—Injection Molding of Conventional Fiber Reinforced Thermosets

Thermoset-based materials with fibrous and/or particulate fillers may conveniently be fabricated by injection molding. The process involves high injection pressures, 50–100 MPa (7.5–15 ksi), and component dimensions are limited by the available mold closing force. Molds are heated to the cure temperature of the material, typically 150–200°C (300–400°F), and the moldings are ejected hot.

These factors combine to limit the process to small- to medium-sized moldings, say, up to 5 kg (11 lb) weight, with an emphasis on much smaller items. Molds and machines are relatively costly in comparison with the requirements for compression or transfer molding; thus, the process would generally be contemplated only when long production runs are envisaged.

When these economic and technical considerations are favorable, the injection molding process is considered to be the most cost-effective method of thermoset fabrication.

3.9.8 Injection Molding of Bulk Molding Compounds (BMCs)

BMCs are fiber reinforced thermoset compounds based on thickened liquid resins such as unsaturated polyester, vinyl ester, and epoxide resins and their derivatives. They are filled with chopped glass fibers with or without additional particulate fillers. The fibers are typically of the order of 5–20 mm (¼–1 in) in length. BMCs are supplied in the form of a viscous dough or paste, often in the form of "rope." Thus, in comparison with the conventional fiber reinforced thermosets discussed in the preceding subsections, BMCs are in paste rather than granular form and contain much longer fibers.

Conventionally, BMCs are compression molded in heated closed molds, as discussed previously. However, they are amenable to injection molding subjected to similar principles as for the conventional thermoset compounds. The essential differences arise from the form of the raw material, the nature of the resin matrix, and the length of the glass fibers. Thus, BMCs are used for larger moldings, the machines are larger, molding pressures are lower, and special feeding devices are necessary.

BMC injection molding is a rapidly developing technology; therefore, it is difficult to set out a general standard for injection molding. The following summary is believed to be representative of current practice.

BMCs are usually based on thickened unsaturated polyester and vinyl ester resins similar to those used in SMC formulations. The cure reaction is based on the free-radical initiated addition polymerization of the monomeric diluent, usually styrene or DAP, that crosslinks the precondensed molecules of the base resin through the unsaturated groups within those molecules. In addition to the thickened resin, BMC contains 20–60% (weight) chopped glass, up to 40% particulate filler (usually some form of calcium carbonate), and often thermoplastics modifiers or low profile additions, together with internal release agents.

The process requires that the BMC be heated to reduce its viscosity and then injected at moderate pressure, 5–10 MPa (750–1,500 psi), into a heated mold. The cure temperature is lower than that of the conventional thermosets, typically 120–150°C (250–300°F).

BMC injection molding machines are similar in principle to those used for thermosets except that they are generally larger, operate at lower pressures, and are equipped with a mechanical feeding device, since the pasty BMC will not self-feed into the throat of the screw. The feeding mechanism is usually a coarse tapered screw set within a casing in the hopper. The pasty compound is picked up at the coarse large diameter end and transported through a convergent passage to the throat of the main injector screw. Simpler plunger operated devices are also used in some machines. Once the material has been introduced to the main screw, it is transported through the water jacketed barrel, where it is heated to 50–70°C. This heat reduces the viscosity of the paste, which is further plasticized by the mechanical action of the screw. (BMCs are generally thixotropic.) The plasticized charge is ultimately injected into the mold by reciprocation of the screw as in the other injection processes described in sections 3.8 and 3.9.

The scale of inhomogeneity of BMCs (i.e., 5–20 mm fiber length) requires that the machine screw be 75 mm (3 in) diameter or greater; otherwise, the long fibers would be severely broken up during passage through the machine. (Note that, in conventional thermosets and thermoplastics, the fiber lengths seldom exceed 2 mm.) This factor has a significant influence on the range of practical machine sizes. The screw design is such that mastication is minimized, and a low L/D ratio and low compression ratio are usual.

Once in the mold, the charge is heated to the cure temperature and retained under pressure until after gelation, when the injection pressure may be relaxed, but the mold is kept closed until cure is substantially complete and the molding rigid enough to be demolded. If necessary, the mold may be opened momentarily at intervals during cure to facilitate the venting of volatiles as in the case of conventional thermosets. Because of the larger size range of BMC moldings in comparison to fiber reinforced thermosets, it follows that cure cycles are correspondingly longer. Cure cycles of 2–5 minutes are common, although development work aims to reduce cure cycle times to 15–30 seconds.

3.9.9 Molds for BMC Injection Molding

Shape requirements are similar to those of other injection molding processes, e.g., adequate draft, no undercuts, and uniform section thickness where possible. However, the less stringent temperature and pressure regimes allow less expensive materials to be used for mold construction. Small molds may be cut from mild steel. It is advantageous to chromium plate or hard-face the cavities and runners if long campaigns are envisaged. Cast iron and cast zinc base or even aluminum alloys may also be used in some instances. The zinc base alloys are particularly useful for large molds since they can be cast close to size in sand molds and then are relatively easy to finish machine to size. Steel reinforcement and heating coils may be cast in situ, and the overall cost of the molds is much lower than in alternative materials.

Venting is again an important consideration in mold design, although BMC cures with less volatiles evolution than the phenolics and similar thermoset materials. However, since BMC moldings tend to be larger, the venting distances are correspondingly longer.

BMC moldings are much more sensitive to fiber orientation induced anisotropy than fiber reinforced thermosets and thermoplastics, and, therefore, the material flow pattern into the mold is much more critical. Irregular flow patterns also result in unsightly swirl patterns on the surface of the molding, which are difficult to eliminate and often show through a subsequently applied paint finish. Runners and gates must be positioned to encourage laminar flow and even filling of the mold cavity. Weld lines should be avoided if possible, although they tend to be less detrimental than with other materials. In general, mold design appears to be less critical than for other injection processes because of the increased scale of BMC moldings with consequent absence of intricate detail, very thin sections, and the need for very fast cycling.

3.9.10 Injection-Compression Molding

Injection-compression molding is a variant of injection molding which is particularly suitable for the larger moldings in BMC and conventional thermosets. The advantage of the process is that the mold closing force requirements are reduced in comparison with straight injection molding, but the advantages of the screw plasticization and delivery of the charge are retained. The principle of operation is that the mold is allowed to retract slightly as the charge is injected and is then closed by the action of the mold closing ram. The molding pressure is thus determined by the mold closing force rather than by the injection pressure. This method allows a larger projected area to be molded for a given press size. In effect, the screw plasticizing train is used merely as a mold charging device. The process is suited to the molding of large area flat moldings without intricate detail that would demand the extra molding pressure of the injection molding process.

Venting is a major problem when making large area moldings; a common practice is to slightly open the mold once or twice during cure of the resin. Ideally, the mold should be opened just before gelation to allow the volatiles to escape before the charge becomes rigid. The subsequent reapplication of pressure finally consolidates the molding.

3.9.11 Summary and Comparison of Methods for Molding Fiber Reinforced Thermosets and BMC

There are three principal processes for molding fiber reinforced thermosets and BMCs, viz., compression, transfer, and injection molding. The choice of process depends on the size and intricacy of the component, the surface finish, and the rate of production requirements.

Compression molding is a simple process needing simple tooling and presses that operate at moderate molding pressures, typically 3–10 MPa (500–1,500 psi) for conventional thermosets and even lower in the case of SMC and BMC. The sizes of compression moldings range from a few grams up to perhaps 5 kg (11 lb) for conventional thermosets, whereas the size of BMC (and SMC) moldings is limited only by the press capacity. For the midpoint of the molding pressure range, i.e., 5 MPa ($\sim$750 psi), 10 kN of press closing force is required for each 20 cm^2 of projected area of molding (2,000 lbf for 2.7 in^2); hence, a 10 MN press is capable of producing moldings of 2 m^2.

Transfer molding operates at somewhat higher pressures, 10–50 MPa (1,500–7,000 psi); consequently, a relatively greater press force capacity is required. The advantage of the process lies in better dimensional tolerances and a more homogeneous plasticization of the compound. Cycle times are also marginally faster. The recent rapid development of the injection molding process for thermosets has removed much of the attraction of the transfer molding process. It is now used mainly for medium to large moldings in conventional thermosets [e.g., 250 g–2.5 kg (9 oz–5.5 lb)]. For smaller thermoset moldings, injection molding would be preferred, whereas for BMC either compression or injection molding could be used.

Injection molding requires still greater pressures of up to 100 MPa (15 ksi) and more. Equipment is therefore massive, especially for large moldings. The process offers the advantages of excellent homogenization of the compound, fast cycling, automated operation, and consistent quality. This process is very widely used for the large-scale production of medium to small fiber reinforced thermoset moldings, and, at lower operating pressures, it is being adapted for large BMC moldings.

In this context the development of modified processes, such as injection-compression molding, is particularly relevant.

3.9.12 Reference

ENGLISH, L. K. "Fabricating the Future with Composite Materials, Part I: The Basics," *Materials Engineering*, 4:15 (September 1987).

SECTION 3.10

*Reinforced Reaction
Injection Molding*

3.10 REINFORCED REACTION INJECTION MOLDING

A. B. ISHAM

3.10.1 General Description

In reaction injection molding (RIM), two highly reactive liquid monomers are carefully metered, brought together in a mixhead, and immediately injected into a heated mold under low pressure. Chemical reaction between the two components produces a cross-linked polymer at the same time the part is being formed in the mold. RIM generally involves the reaction of a liquid polyol and isocyanate to form a urethane polymer. This mold filling and polymerization takes only a few seconds. The entire process, including curing and demolding, takes from 1 to 5 minutes depending on the resin, the part thickness, and the capabilities of the processing equipment. A 2 to 3 minute cycle time can be expected in most cases, however.

There are several basic differences between RIM and other injection or compression molding processes. Because of the low pressure of the injection, RIM requires less expensive molds, little heat input, and low locking forces. RIM processing also offers low capital cost, low energy use, advantages of automation, and fast molding cycles.

The basic RIM process requires temperature-controlled storage tanks, metering pumps, a mixer, temperature-controlled molds with a mold clamping system, and associated piping, hoses, filters, and controls. Very detailed descriptions of all aspects of RIM technology can be found in two books on the subject [1,2].

Despite the advantages of the process, RIM urethane materials are sometimes limited in their applications because of their high coefficient of thermal expansion and poor resistance to sagging or slumping at elevated temperatures. However, by altering the RIM process to add reinforcements or fillers to the urethane, these properties can be improved and the application of the material extended, while retaining the processing advantages. With reinforced reaction injection molding (RRIM), parts are more resistant to temperature changes than those from unreinforced RIM. RRIM also provides large, complex parts in economic cycle times.

3.10.2 Choice of Materials

A description of the basic polyols and isocyanates used in RIM is covered in section 3.3. Most suppliers offer several different urethane intermediates to the molding industry. The biggest differences noted are in the flexural modulus and the maximum tensile elongation of the final product. By varying the basic polyol and chain extender in the system, the flexural modulus can be varied from about 172 to 1,700 MPa (25 to 200 ksi) and the maximum elongation from 250 to 50%, respectively. The isocyanate most commonly used is diphenylmethane 4,4'-diisocyanate (MDI).

Some RIM systems incorporate a small amount of a fluorocarbon blowing agent in order to help create a microcellular structure in the core of the molded part. The blowing agents are not used to create a low density foam but, rather, to provide a slight density reduction while improving the packing of the mold. Generally, if blowing agents are used, the density of the urethane is reduced to about 1.0 g/cm^3.

The reinforcement in the RRIM material system can be added to either or both of the RIM components prior to mixing, but it is normally blended only with the polyol. A variety of reinforcing materials has been used with varying degrees of success. Milled glass fibers are most commonly used and presently are the product of choice. Chopped glass fibers, wollastonite, carbon fibers, and mica flakes are other reinforcing materials for RIM urethanes. Fillers, such as chalk, wood flour, milled rubber scrap, urethane foam scrap, and barium sulfate, also have been used in RIM. In the final composite, the advantages of fiber reinforcements over mica are better flexural modulus, sag properties, and coefficient of thermal expansion, in addition to less reduction of elongation and impact strength. Milled glass fibers are preferred over chopped strands because of their simple processing, their capability for higher loadings, and the improved surface quality of finished parts [3]. Combinations of fibers and particulate fillers are considered in the development of new RRIM systems.

Raw Material Storage

The RRIM manufacturer may either purchase basic ingredients and formulate them in-house or buy fully or partially compounded formulations. The economics affecting these choices are based on his production volumes and technical capabilities. In either case, proper storage and handling to protect the materials from contamination degradation are vital [1,4].

Polyols. Polyols should be stored in mild steel tanks that are clean and rust-free to prevent contamination by foreign substances. Contact with copper or copper alloys in either tanks or piping should be avoided.

Exposure of polyols to moisture can produce undesirable gas formation during the gelling reaction. Moisture contamination can be prevented by topping the storage tanks with a blanket of dry gas, preferably nitrogen. Polyols, which normally have moderately low pour temperatures, can become viscous at cool temperatures; therefore, tank temperature should be kept between 21 and 43°C (70 and 110°F). Higher temperatures will degrade the material.

Isocyanates. Isocyanates can be stored in mild carbon steel or stainless steel tanks, although stainless steel may be somewhat expensive for this application. If product discoloration must be prevented, tanks should be lined with phenolic, epoxy, or urethane materials. Contact with brass or aluminum should be avoided. Storage temperatures should range between 24 and 30°C (75 and 100°F) depending on the specific material. It is important to maintain the ingredients (MDI and MDI prepolymers) in a liquid state to insure ease of processing and of transfer.

Preventing moisture contamination is more important with isocyanates than with polyols. Water will react with the isocyanate to form a hard precipitate and reduce isocyanate content, which can adversely affect both the ease of processing and final material properties. To keep isocyanates moisture free, they should be stored in a dry, non-reactive atmosphere at a slight positive pressure. All transfer lines should be plugged or capped when not in use. To prevent moisture contamination when isocyanates are discharged from drums, the vent should be equipped with a dry nitrogen or dry-air "breather." Drums can be unloaded with air-driven or hand-operated pumps, and valves should be capped when not in use. Isocyanate removed from a drum should be replaced with dry nitrogen, or dry air, containing no more than 65 ppm of water.

Other Ingredients. Fluorocarbon blowing agents are generally stored in the drums in which they are received. If the volume of use requires tank truck delivery, then storage tanks, normally steel, will be required. Either steel or copper can be used in piping systems. Information on minimum storage tank design pressure should be obtained from the fluorocarbon supplier. If catalysts are to be stored, suppliers should be contacted for instructions. All reinforcement materials should be kept in a dry, covered area.

Safety

Every manufacturer of chemicals for the RIM urethane formulation supplies extensive recommendations for the safe handling of its products. These recommendations should be thoroughly understood by the purchaser and followed to the letter.

Since isocyanate vapors represent a potential hazard to personnel, precautions must be taken against unacceptable vapor concentrations. As defined by OSHA, the limit of isocyanate exposure is 0.02 ppm of air; proposed legislation would reduce this limit to 0.005 ppm. Because, under normal use conditions, the MDI-based isocyanates used in RRIM formulations have low vapor pressures, they can be handled safely with adequate ventilation. Good ventilation is particularly important at unloading points, mixing and molding stations, and waste disposal areas. Special measures need to be taken to prevent worker exposure to vapors at elevated temperatures or to vapors resulting from massive spills. Spills should be covered with an absorbent material and neutralized with a dilute solution of aqueous ammonia and glycerine or ethylene glycol.

The polyols and fluorocarbon blowing agents are relatively safe materials that require little special care. Fluorocarbons will decompose to acidic toxic materials in the presence of flame or high temperature, but there should be little hazard in normal RRIM processing.

3.10.3 The Molding Process

Basic RIM equipment can be used for RRIM. However, the addition of a reinforcement does affect certain aspects of the RIM process; there are some substantial differences between the two processes. In RRIM, the reinforcement normally is premixed with a polyol, which is combined with isocyanate in a conventional high-pressure mixhead and then injected into a closed mold under low pressure. Certain changes in RIM equipment are required because the fiber filled materials, which have higher viscosities than unfilled systems, are much more abrasive. Viscosities of reinforced urethanes range up to about 6 Pa·s (6,000 cps) compared to a maximum of 2 Pa·s (2,000 cps) for unreinforced compounds. As a result of high viscosity and the abrasiveness of the slurries, the metering pumps used to measure components into the mixhead in RIM processing are not applicable to reinforced systems. For RRIM, therefore, the pumps are replaced by high-pressure metering cylinders. A simple schematic of the RRIM process is shown in Figure 3.10-1.

Other changes required in RIM equipment for RRIM processing include [5] the following:

1. Use of wear resistant material for parts exposed to the abrasive reinforced slurry
2. Use of material feed lines that are as straight and short as possible to eliminate clogging and fiber settling

In all urethane equipment, some provision is made for recirculation of the liquids and slurries throughout the system to eliminate material stratification or fiber settling.

In the following brief description of the RRIM process, the basic processing steps—addition of the reinforcement, metering, mixing, molding, and demolding—are considered.

Addition of Reinforcement

Milled glass fibers, or other reinforcement materials, are first blended with components in the holding tanks, generally with the polyol. The reinforcement must be

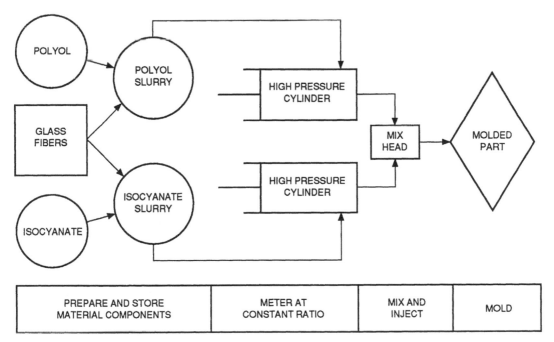

FIGURE 3.10-1. Schematic of reaction injection molding with glass fiber reinforced urethanes [6].

added before the urethane materials are mixed together because of the rapid reaction that occurs between the ingredients after mixing. The reinforcing material is added in carefully measured amounts to obtain a final compound with the desired volume fractions.

Reinforcement loadings must be monitored carefully to prevent a slurry that is too thick for processing. Viscosity of the slurry with glass-fiber reinforcement depends on average length of the filaments. Common polyol slurries contain 20 to 40% of glass fibers that have been hammermilled through a 1.5 mm ($^1/_{16}$ in) screen. This slurry, when combined with an unfilled isocyanate, will produce a RRIM composite containing from 10 to 20% reinforcement by weight. Some equipment manufacturers claim that their systems can handle slurries with greater than 50% loadings [7].

Once the reinforcement is added to the tank, several types of mixers are suitable for blending the fibers and liquid components into a thick slurry. A low shear mixer with plow-shaped mixing tools, for example, has been used to prepare excellent slurries in a short time. Mixers with heating jackets and with vacuum capability are also available. Such equipment can be used to control viscosity and to remove entrapped air bubbles in the slurry.

After preparation, the reinforced slurry is piped to a "day tank" for subsequent transfer to the mixhead in measured "shots." In the day tank, it is slowly and continuously agitated by mixing impellers to prevent settling of the fibers. These tanks are pressurized at 100 to 410 kPa (15 to 60 psi), depending on the urethane system, and have heated jackets for temperature control. Unfilled components are stored in a similar manner prior to the injection.

Metering the Components

Injection in RIM processing is accomplished by high-pressure metering cylinders, generally in conjunction with low-pressure "slurry" or feed pumps. Conventional RIM machines with high-pressure pumps can be converted to RRIM processing by the addition of a "lance" or "dosing" cylinder, which will change the function of a metering pump from one of metering a urethane chemical to that of metering hydraulic fluid which powers the cylinder.

To move the slurries to the mixhead, the metering cylinders are first filled with material from the day tanks. The slurry can be moved directly from the pres-

surized tanks by air pressure but, more often, is transferred with an assist by feeder pumps. The feeder pumps located between the day tanks and the metering cylinder are designed to handle slurries under pressures up to 1 MPa (150 psi). Feeder pumps can provide continuous recirculation, but if the metering cylinders are filled directly from pressurized day tanks, some recirculation can be obtained by stroking the cylinder at low pressure and allowing the slurry to return to the day tank. In order to maintain pressure and prevent clogging in the metering cylinders and the mixhead, bends and elbows in the lines should be kept to a minimum.

Mixing

After the metering cylinder is filled, it is driven forward by hydraulic pressure to deliver slurry to the mixhead at a known rate. The other component is injected from another cylinder at the same time. Delivery of the two components must be closely synchronized in order to ensure a uniform reaction and consistent properties in the cured product.

In some equipment, during the first part of plunger travel, the valve to the mixing chamber is closed, and the slurry is recirculated through the head into the return lines and back to the day tanks. After this preliminary recirculation, the mixhead valve is opened, and the slurry enters the chamber to be mixed and delivered to the mold. After the metering shot, the mixhead valve is closed, and the slurry recirculates back into the day tanks.

The mixhead is designed to develop turbulence in the mix chamber to intimately mix the two components. Turbulence is created by stream impingement at high pressure, generally between 10 and 21 MPa (1.5 and 3 ksi).

For impingement mixing, the mix chamber is normally a small cylinder with openings that allow the slurries to enter from opposite sides of the chamber. The quality of impingement mixing depends on the Reynolds number and momentum of each stream. For best mixing, the streams should have equal momentum at the time they meet, and the Reynolds numbers should be above a critical minimum; however, that minimum has not been clearly defined. With unreinforced RIM components, Reynolds numbers over the range of 500 to 4,000 have been evaluated [8].

For RRIM processing, standard RIM mixheads are redesigned to withstand the abrasiveness of the slurries.

The Molding Process 185

This involves use of a hard material — tungsten carbide, for example — for nozzles, replaceable inserts, valve orifices, and any other parts that handle the high-pressure slurry streams.

Molding

The compound, which is cleared from the mixing chamber by a close-fitting ram, flows directly to the mold. It is common practice in urethane molding to add an aftermixer between the mixhead and the mold cavity to insure that there are no slugs or fluids at improper ratios that will not set up properly. This aftermixer is normally designed into the mold itself. The mold is normally filled from the bottom so that air can easily push out ahead of the flow. The shot fills the mold to about 90%, and expansion during chemical reaction of the polyol and isocyanate completes the fill. The compounds react and gel within 2 to 10 seconds from the start of injection. The mold must remain closed long enough to provide a part with adequate "green strength," i.e., enough cure to allow the part to be removed and handled without damage. Depending on the resin, the cure time is generally between one-half minute and one and one-half minutes. RRIM parts will continue to gain some strength and toughness after removal from the mold. Normally, for most formulations, parts are postcured in an oven for about one hour at temperatures from 120 to 160°C (250 to 320°F).

Demolding

Once the mold has opened, the operator has several options for removing the part, depending on part size and design. In an ideal situation, the part can be gently pulled from the mold. Knock-out pins, automatic slides, or pneumatic devices in the mold are sometimes used for demolding assistance. For small flexible parts, the operator may use a rubber spatula to pry the part loose, or a flexible part can often be removed by inserting an air nozzle beneath a corner and blowing it off.

Part removal can distort the part slightly. The capability of the part to recover will depend on precise control of metering ratios, formulation reactivity, and degree of cure [1].

Mold release compounds, applied before molding, also facilitate release; they may be wax compounds, metal soaps, silicone coatings, or internal mold release compounds. Wax compounds have proven most suc-

cessful and can be easily removed from the part by conventional degreasing operations. Silicone lubricants provide excellent separation but form a coating on the part that is nearly impossible to remove. The same problem arises with the internal mold release compounds. The major problem with metal soaps is the catalytic effect on the polymerization reaction, which can cause surface defects in the part.

After the part is removed, the mold must be cleaned to prevent release compound buildup; it is then hand sprayed or brushed with release compound in preparation for the next shot.

Mold Design and Construction

Proper mold design goes hand in hand with part design. The overall design of the part should [2]

1. Allow uniform filling of the cavity
2. Permit relatively nonrestricted expansion if the formulation contains a blowing agent
3. Include draft angles large enough to facilitate demolding
4. Allow the part to fall free when the mold is opened or to stick preferentially to the back of the contour

The mold itself must accommodate these design features and must be stiff enough to withstand molding pressures and to maintain structural integrity during opening and part demolding. Since molding pressures are low, a variety of materials can be used to construct the molds; aluminum, steel, and nickel are used most commonly and provide good surface finish, good temperature control, resistance to molding pressures, and tight sealing. Kirksite (zinc alloy) and epoxy have been used as prototype mold materials but are not suitable for production applications.

The path from the mixhead to the mold cavity is provided by gating devices and filling sprues which are built into the mold. Entry into the mold should be along the mold parting line, and the location should provide the shortest possible distances from the point of entry to the furthest points of the mold cavity. The opening into the mold cavity should have a much smaller cross-sectional area than the fill path itself. This constriction creates a pressure to force air from the sprue into the mold cavity and prevents bubble formation caused by air entrapment in the sprue system. The design should insure uninterrupted and nonturbulent flow of the mixed urethane system into the mold. Steady flow will insure that air is forced out of the mold ahead of the compound. For high through-put, cylindrical-shaped runners that guide the compound along the mold cavity in a channel with a rectangular or semicircular cross section are used.

Contact surfaces of RRIM molds should prevent escape of liquid but not of gas. Air entrapment in the mold is undesirable; it leads to bubbles near the surface that may be visible. Thus, the design should provide for venting at the contact surfaces at the highest point of the mold. Vent holes may also be placed in bosses and projections. Tilting the mold improves venting by forcing entrapped air to a high point in the cavity from where it may escape. Mold contact faces should be flat, and pressure should be applied uniformly around the perimeter during injection. The width of these faces will vary with the mold material; e.g., 6 to 12 mm (¼ to ½ in) is recommended for steel, 12 mm (½ in) for nickel, and 12 to 18 mm (½ to ¾ in) for aluminum [7].

The mold should also be designed for easy release of the part, with draft angles of no less than 1.5 degrees for smooth finish parts and 3 degrees for rougher parts. Undercuts and negative draft angles should be avoided as much as possible, although minor negative draft angles may be used to make the part stick to one mold face. Where undercuts cannot be avoided, either removable or permanent inserts may be included in the design to assist in part release. These inserts can be manually placed for each molding or can be activated with air cylinders. Permanent inserts may be used to accept fastening devices or as structural components of the part.

The mold also provides thermal control of the exotherm to control polymerization. The high reactivity of RRIM formulations produces an exotherm that is largely retained in the part and effectively cures the foam core. However, material at the part surface can gain or lose heat through contact with the mold surface. To control this heat transfer, coils in the mold maintain a mold temperature near the reaction temperature, generally 50 to 60°C (120 to 140°F).

Mold cooling coils should be spaced to keep the cavity surface and, thus, the part surface at a uniform temperature to help insure uniform skin thickness. Connecting two or three sets of coils in parallel rather than having a single coil in each half of the mold is beneficial to part quality. High-velocity fluid flow in the coils enhances heat transfer and improves quality.

Mold Clamping and Support

Filling of the mold and proper expansion are facilitated by proper position of the mold in supporting equipment. The molds can be supported in a variety of presses or other devices, which also apply pressure to hold the mold halves together during injection, provide means to open the mold, and usually assist in part removal. For RRIM applications, molds are generally supported in universal presses, automotive presses, or limited movement presses. The universal presses can move the mold in three axes for easy injection, proper fill, and easy self-removal. Automotive presses provide only X-Y tilt to direct fill and expansion flow. Limited movement presses are less costly and can be used to meet large production requirements of several parts with the same general size and shape [1,7].

Although molding pressures for RRIM products are quite low—seldom greater than 350 kPa (50 psi)—large

Table 3.10-1. Thermal properties of RIM urethanes reinforced with milled glass fibers [6].

Property	Units	Urethane System and Glass Content, %					
		Low Modulus		High Modulus		Rigid	
		0	15	0	15	0	15
Coefficient of	10^{-6}/°C	158	86	128	65	108	63
Thermal Expansion	(10^{-6}/°F)	(88)	(48)	(71)	(36)	(60)	(35)
Cantilever Sag[1]	cm	1.1	0.2	0.5	0.3	4.6	1.0
in Oven, 1 hr	(in)	(0.4)	(0.1)	(0.2)	(0.1)	(1.8)	(0.4)
Mold Shrinkage	%			1.2	0.6	2.1	0.5

[1]For low and high modulus systems: 10 cm beam @ 125°C (4 in beam @ 258°F).
For rigid systems: 20 cm beam @ 163°C (8 in beam @ 325°F).

parts require clamping forces of several tons; consequently, they require large presses. Also, the mold support must have sufficient hydraulic capability to open the mold against the adhesive characteristics of most urethane systems. This "break-away" force may be several times greater than the clamping force [2].

In addition to the mold clamp itself, the molding facility for RRIM may include the following:

1. Conveyorized mold devices having a fixed delivery system and moving molds
2. Carousels with a fixed delivery system and a mold fixed to a moving carrier
3. Turreted systems with a fixed mold and mold carrier and a delivery system that moves from mold to mold. Such systems are usually used with smaller parts.

Safety

OSHA regulations and safety requirements for reaction injection molding apply equally to RRIM. In addition, the mold should be vented to the atmosphere during filling, and exhaust ventilation should be provided to protect the operator from any possible contact with isocyanate vapors.

3.10.4 Mechanical Properties

The addition of reinforcement to RIM materials improves the stiffness (as measured by flexural modulus), the coefficient of thermal expansion, and the resistance to sag, while reducing the maximum tensile elongation and the impact strength. Exact properties are, of course, influenced by the urethane system, the reinforcement material, the material characteristics (e.g., milled fiber length), and the reinforcement loading.

In one assessment of RRIM property data, urethane systems were classified by their flexural modulus [6]:

1. Low modulus—flexible urethanes with a modulus between 140 and 280 MPa (20 and 40 ksi)
2. High modulus systems—semirigid urethanes with a modulus between 690 and 1,030 MPa (100 and 150 ksi)
3. Rigid systems—highly rigid urethanes with a modulus of 1,720 to 2,070 MPa (250 to 300 ksi)

Table 3.10-1 shows thermal properties of these three

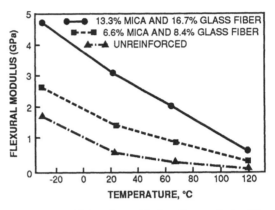
FIGURE 3.10.-2. RIM urethane with glass fibers and mica [11].

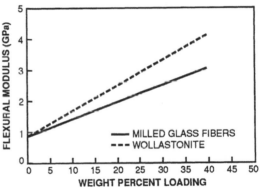
FIGURE 3.10-3. Reinforced high modulus urethane [13].

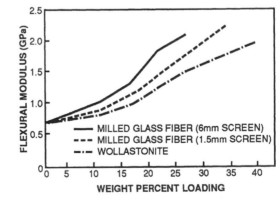
FIGURE 3.10-4. Reinforced high modulus urethane [14].

Table 3.10-2. Mechanical properties of RIM urethanes reinforced with milled glass fibers [6].

| | | Urethane System and Glass Content, % | | | | | |
| | | Low Modulus | | High Modulus | | Rigid | |
Flexural	Units	0	15	0	15	0	15
Flexural	GPa	0.20	0.36	0.83	1.47	1.79	297
Modulus	(ksi)	(29)	(52)	(121)	(213)	(260)	(430)
Tensile	MPa	19.4	17.9	31.1	33.8	53.0	60.1
Strength	(ksi)	(2.8)	(2.6)	(4.5)	(4.9)	(7.7)	(8.7)
Tensile Elongation	%	123	63	83	27	16	11

Table 3.10-3. Directional properties of a RIM urethane reinforced with milled fibers [9].

| | | | Glass Fiber Content, % | | | |
Property	Units	Direction[1]	0	5	10	15
Coefficient of	$10^{-6}/°C$	M	185	100	55	40
Thermal Expansion		T	185	185	175	165
	$(10^{-6}/°F)$	M	(103)	(56)	(31)	(22)
		T	(103)	(103)	(97)	(92)
Tensile Strength	MPa	M	29	25	23	23
		T	27	24	22	20
	(ksi)	M	(4.2)	(3.6)	(3.3)	(3.3)
		T	(3.9)	(3.5)	(3.2)	(2.9)

[1]M = Flow direction.
 T = Transverse to flow direction.

Table 3.10-4. Properties of a RIM urethane, reinforced with glass fibers and mica [10].

| | | Reinforcement, 15% by Weight | | | |
| | | | Milled Glass Fibers | | |
Property	Units	Unreinforced	1.5 mm (¹⁄₁₆ in) Screen	3 mm (⅛ in) Screen	Mica Flake
Density	g/cm³	1.0	1.27	1.29	1.30
Flexural	GPa	1.21	2.17	2.50	2.14
Modulus	(ksi)	(175)	(315)	(363)	(310)
Tensile Elongation	%	95	46	22	10
Coefficient of	$10^{-6}/°C$	81	36	18	34
Thermal Expansion	$(10^{-6}/°F)$	(45)	(20)	(10)	(19)
Notched Izod Impact	J/m	373	62	165	63
	(ft-lb/in)	(7.0)	(1.7)	(3.1)	(1.8)

systems, both unreinforced and with 15% glass-fiber reinforcement. The table shows that glass-fiber reinforcement improves several properties in urethanes:

Polymerization shrinkage is reduced. This characteristic allows for the production of molded parts that have a size and shape closer to the mold dimensions than unreinforced RIM parts.

Thermal expansion is reduced. The coefficient of thermal expansion is a measurement of dimensional stability over a wide temperature range. Because variation in the temperature of glass reinforced systems will cause less dimensional change, molded parts can be designed with closer tolerances. A lower thermal expansion also insures that less movement will occur in a changing temperature environment when glass reinforced urethanes are attached to metal supporting structures.

Droop and sag at elevated temperatures are reduced. While being painted, urethane parts require fixtures to help maintain their shape in the high temperature environment of the paint bake oven. A lower level of droop reduces the complexity and cost of the fixtures. In service, the lower level of droop helps prevent sagging or distortion when the urethane part is clamped along one edge.

Table 3.10-2 shows mechanical properties for the same systems as in Table 3.10-1. These properties are influenced by both the glass fibers and the urethane chemistry. Thus, it should be possible to formulate different RRIM material systems to cover a wide range of strengths and stiffnesses. As a rule of thumb, however, increased stiffness achieved by either chemical or reinforcement means generally results in reduced tensile elongation.

As a rule of thumb, mechanical properties follow a linear relation to the amount of reinforcement over the range of 0 to 25%. Table 3.10-3 shows data for polyurethane with reinforcement loadings of 5, 10, and 15%. This table also indicates anisotropy in the coefficient of thermal expansion due to the orientation of the glass fibers that can occur from flow into the mold. At 15% glass-fiber content, the coefficient dropped to one-fourth of its initial value in the direction of flow, whereas very little change occurred in the transverse directions. In this study [9], the tensile strength was largely unaffected by the orientation of the glass fibers.

The effects of fiber length on properties are shown in Table 3.10-4 [10]. These data indicate that 3 mm (⅛ in) milled glass reinforced materials have improved properties over those reinforced with 1.5 mm (¹/₁₆ in) milled glass. Properties of mica reinforced urethane are comparable to those of the 1.5 mm (¹/₁₆ in) glass reinforced product. The table also shows the significant reduction in impact strength of reinforced compounds. Figure 3.10-2 shows flexural modulus curves for mica and glass-fiber combinations compared with unreinforced material [11]. Mica has also been used in combination with small amounts of carbon fibers for improved properties [12].

There is considerable interest in using milled glass fibers along with the mineral wollastonite. Figure 3.10-3 indicates that higher flexural modulus can be obtained with wollastonite than with milled glass fibers [13]. However, another study showed better stiffness for milled glass fibers, as seen in Figure 3.10-4 [14]. This latter study concluded that wollastonite is less effective at improving properties of RRIM urethanes than either 1.5 or 6 mm (¹/₁₆ or ¼ in) milled glass fibers at equal loadings.

To be effective, any process must be able to provide improved performance at reasonable cost. Applications for RRIM develop rapidly, as improved chemical systems and processes evolve [15,16].

3.10.5 References

1. BECKER, W. E. *Reaction Injection Molding.* Van Nostrand Reinhold Co., New York (1979).
2. SWEENEY, F. M. *Introduction to Reaction Injection Molding.* Technomic Publishing Co., Inc., Lancaster, PA (1979).
3. SEEL, K. and L. Klier. "Processing and Properties of Reinforced Bayflex," *Proceedings of the SPE National Technical Conf.*, Detroit, Michigan, pp. 117–118 (November 1979).
4. HAWKER, L. E. "Follow These Guidelines on Storing and Handling RIM Materials," *Plastics Technology*, 25(11):69–71 (October 1979).
5. MIKULEC, M. J. "RRIM—A New Process for the Automotive Industry," *Proceedings of the 34th Annual Technical Conference of the Reinforced Plastics/Composites Institute*, New Orleans, Louisiana, Section 11-D (February 1979).
6. ISHAM, A. B. "Reaction Injection Molding with Glass Fiber Reinforcement," Society of Automotive Engineers Technical Paper Series, No. 780354, Detroit, Michigan (February 1978).
7. SISTAK, B. J. "RRIM Opens New Doors for Processing Polyurethanes," *Plastics Design and Processing*, 19(10): 20–26 (October 1979).

8. MALGUARNERA, S. C. and N. P. Suh. "Liquid Injection Molding, I. An Investigation of Impingement Mixing," *Polymer Engineering and Science*, 17(2):111–115 (February 1977).

9. SCHULTE, K. W., et al. "Glass Reinforced RIM Polyurethanes—Properties and Processing Techniques," *European Journal of Cellular Plastics*, 2(2):61–67 (April 1979).

10. MACGREGOR, C. J. and R. A. Parker. "Controlling RIM Properties with Reinforcements," *Plastics Compounding*, 2(5):53–57 (September/October 1979).

11. "Reinforced Reaction Injection Moulding—Friend or Foe to SMC," *Reinforced Plastics*, 23(10):324–330 (October 1979).

12. "Improving RIM Technology Multiplies the Options," *Modern Plastics*, 55(2):34–37 (February 1978).

13. "RIM Emerges as a Major Process," *Modern Plastics*, 55(8):37–39 (August 1978).

14. KOSTECKI, R. J. "Selection of Reinforcements for the Reaction Injection Molding Process," *Proceedings of the SPE National Technical Conference*, Detroit, Michigan, pp. 14–17 (November 1979).

15. BRACE, D. R. "Structural RIM—The Material and Process Technology for High Speed Specialty Composite Production," *Proc. from Advanced Composites—The Latest Developments, ASM International*, pp. 159–162 (1986).

16. ECKLER, J. H. and T. C. Wilkinson. "Processing and Designing Parts Using Structural Reaction Injection Molding," *Journal of Materials Shaping Technology*, 5(1):17–21 (1987).

SECTION 3.11

Filament Winding

3.11.1 Equipment and Winding Patterns

Introduction and History

The modern practice of filament winding began about 1947 in the form of a government contract with M. W. Kellogg Company to develop a lightweight pressure vessel. In concept, however, filament winding is not that new. In the Middle Ages cannons were often made from iron staves that were overwrapped with steel wire. During World War II a similar technique was used for high-pressure oxygen bottles that were made from aluminum overwrapped with steel wire. The steel wire minimized the likelihood of an explosive failure in the event of a projectile striking the bottle. The advent of continuous glass fibers (developed in the late 1930s) brought a new dimension to the practice of filament winding. Another strong contributing factor to the success of the filament winding technology was the development of liquid thermosetting resins, such as polyester and epoxy.

During the 1950s the technology of filament winding was given increased emphasis because of the need for lightweight pressure containers for the development of rockets for missiles. One of the earliest missiles, the Polaris A1—a two-stage, submarine-launched ballistic missile—had a range of 1,900 km (1,200 miles) using a steel container [1]. The A2 Polaris had a steel first stage and a filament wound fiberglass second stage that increased its range to 2,400 km (1,500 miles). In the A3 Polaris missile both stages used filament wound cases. In addition to a lighter weight, the A3, which went into operation in early 1964, used a hotter propellant. These combinations enabled an increase in the A3's range to 4,000 km (2,500 miles).

Since that beginning, virtually all solid propellant missiles have used filament wound rocket cases for the propulsion system. The list includes Minuteman, Poseidon, and Trident. In recent years, Kevlar (an organic fiber) has supplanted fiberglass due to its superior strength-to-weight ratio.

A large amount of effort was expended in government programs dealing with winding technology, fiberglass sizings, epoxy resins, structural designs, and stress analysis methods [2]. Much of this technology has been of great benefit to the commercial filament winding industry even though most commercial filament windings are based on polyester resins rather than epoxies.

Commercial filament winding began in the mid-to-late fifties, and by 1963 the number of commercial filament wound products exceeded that of military items [3]. The first products were filament wound pipes and tanks, which are still the largest production items by far. The impetus came from the economics of fiberglass composite structures, as compared to corrosion resistant metal alloys or lined steel. The filament wound structure was less costly than stainless steel and gave superior performance in environments such as hydrochloric acid [4].

In the electrical industry the combination of high strength with electrical resistance encouraged the development of filament wound insulating structures such as high voltage fuse tubes, circuit breakers, and transformer cases. In recent years tapered lighting poles have been developed [5]. Other filament wound electrical products (often in combination with pultrusion) have included truck-mounted booms for aerial platforms and hotline tool handles [6].

Winding Patterns and Determination of Physical Properties

There are two types of patterns normally used in filament winding: helical winding, in which a constant angle is maintained, and biaxial winding, where two or more separate winding angles (usually 0° and 90° to the rotational axis) are used.

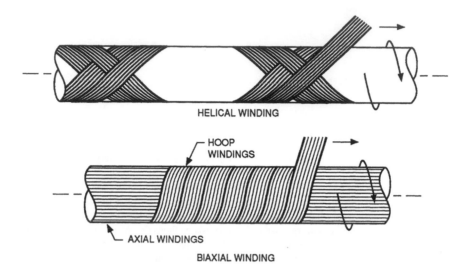

HELICAL WINDING

BIAXIAL WINDING

FIGURE 3.11-1. Types of filament winding patterns.

The sketches shown in Figure 3.11-1 illustrate the two winding patterns. Helical winding is characterized by a plus and minus winding angle (as measured from the rotational axis), giving a balanced structure with circumferential and longitudinal properties dependent upon the winding angle used. A 45° winding angle gives equal properties in the axial and hoop directions. Biaxial winding may have hoop windings and axial wraps or another combination of low angle and high angle windings.

A recent development using transverse filament tapes [7] also achieves a biaxial structure by filament winding. The transverse filament strands are stitched together with polyester yarns; in the winding process the transverse tapes are overlapped to achieve bonding. The principal application is in large 45 m (150 ft) long blades for energy-producing windmill blades.

Helical winding is more common because the processing time is ordinarily less; however, biaxial patterns enable special properties to be obtained and, for closed end structures, have a processing time advantage. In combination with either type of winding pattern, other materials may be a part of the completed structure such as

1. Mats or woven cloth
2. Chopped fiber spray-up
3. Thermoplastic or other liners
4. Metal liners or inserts

Richard Young (often called the father of filament winding) first developed the equations for "netting analysis" for the helically wound structures first used for open or closed end cylinders [8,9]. The netting analysis assumes that there is no contribution from the matrix portion of the composite structure and that all fibers are tensioned equally. Thus, there is no transfer of stress from fiber to fiber; hence, the entire structure must be in tension. The sketches and equations in Figure 3.11-2 illustrate the principles of Young's netting analysis.

Assume the following:

S = Unit strength of parallel filament band
S_H = Unit strength of cylinder in hoop direction
S_A = Unit strength of cylinder in axial direction
t = Wall thickness
α = Wind angle
W = Band width
L = $W/\sin \alpha$
F_1 = Total strength of filament band in α direction
F_2 = Total strength of part in the axial (hoop) direction

$$F_1 = \frac{SWt}{2}$$

$$F_2 = \frac{\sin \alpha \; SWt}{2}$$

$$S_H = \frac{2F_2}{Lt} = \frac{2 \sin \alpha \; SWt}{2W(\sin \alpha)^{-1}t} = S \sin \alpha$$

Similarly, $S_A = S \cos \alpha$. Therefore, for a cylinder with a 2:1 strength ratio (2 hoop:1 axial):

$$\frac{S_H}{S_A} = \frac{2}{1} = \frac{S \sin \alpha}{S \cos \alpha} = \tan^2 \alpha$$

$$= \text{arc tan } \sqrt{2} = 54.75°$$

More realistic analyses [10–12] allow the estimation of the contributions of both fibers and matrices on the finished laminated structure (see also Volume 2).

Filament wound structures are normally designed to carry loads through balanced stresses in the windings. Difficulties are often created when openings, holes, or ports have to be included in the filament wound structure.

When the filaments are cut by a hole in the structure, the primary tension load is transferred into the matrix, which will then transfer the load to the uncut filaments at the sides of the hole. Depending upon the size of the hole and the load level, this shear force may exceed the strength of the resin and create a failure condition around the hole. Parady in 1965 [13] discussed the problems of reinforcing holes in filament wound structures and offered several solutions:

1. Locate the openings in areas where there is the least likelihood of a shear failure.
2. Avoid cutting any filaments by displacing the glass

filaments around the opening (this can sometimes be done by installing a cone on the mandrel during the winding process).
3. Add either a filament wound mat or windings where the holes are cut.
4. Add glass cloth tape or filament wound washer type reinforcements.

For commercial structures such as pipes or tanks, a common method is to use combinations of cloth and mat to reinforce around an opening such as a nozzle or manway. ASTM D3299 [14] specifies reinforcement of the tank wall for the attachment of nozzles. This specification requires that added reinforcement shall extend to one diameter of the opening away from its center. The thickness of the added reinforcement is essentially the same as that of the tank wall with a minimum thickness of 4.8 mm (3/16 in). The construction and reinforcements are not specified in the ASTM Standard for the tank wall buildup, but most producers use alternate layers of mat and woven roving in areas of high shear stress.

Local reinforcements are also required for attachments to filament wound structures, for joints between pipe sections, and when a sudden change of direction is required in a structure—such as for ribs or external attachments. Various types of cloth and mat are often used for these types of added reinforcements.

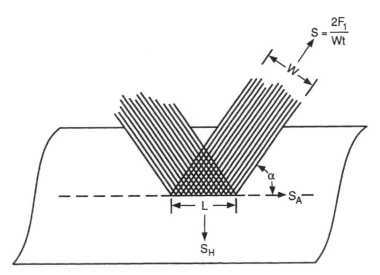

FIGURE 3.11-2. Netting analysis derivation.

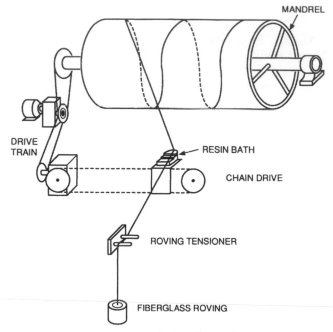

MANDREL

DRIVE
TRAIN

RESIN BATH

CHAIN DRIVE

ROVING TENSIONER

FIBERGLASS ROVING

FIGURE 3.11-3. Helical winding machine.

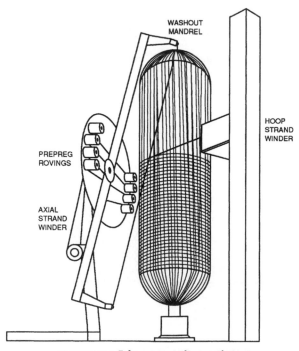

WASHOUT
MANDREL

HOOP
STRAND
WINDER

PREPREG
ROVINGS

AXIAL
STRAND
WINDER

FIGURE 3.11-4. Polar wrap winding machine.

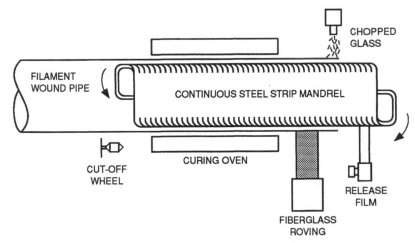

FIGURE 3.11-5. Drosthom continuous winder.

Winding Equipment and Accessories

In the beginning of the filament winding industry (as in the case of most new process developments) all equipment was custom-made. This custom-made equipment was usually satisfactory but frequently expensive and often disappointing in flexibility. Since about 1965, winding equipment has become available from a number of machine manufacturers.

The most widely used and most versatile machine is the horizontal helical winding machine that holds the mandrel horizontal and has a carriage that moves back and forth parallel to the long axis of the mandrel. Figure 3.11-3 illustrates the principles involved. A different type of machine, which was developed primarily for missile cases, is the polar wrap winding machine (see Figure 3.11-4). This winder holds the mandrel in a vertical position and swings a rotating arm to lay down fibers in a polar wrap almost parallel to the long axis of the mandrel. During each wrap the mandrel rotates one band width. Hoop wraps are wound on separately. A third machine type, which has come into use since about 1970 (Figure 3.11-5), is a continuous filament winding machine that is primarily used for making long lengths of large diameter pipe.

In selecting filament winding equipment, it is highly desirable to provide for a wide range of ratios between carriage travel and mandrel revolutions. The horizontal helical winder fulfills this condition best. At high angle (circumferential) wraps, the carriage will travel very

slowly, moving only a small amount for each turn of the mandrel. At the other extreme, with low angle windings, the carriage may move nearly the full length of the mandrel for each revolution.

In addition, the ribbon may be positioned precisely next to the band that was laid down previously, so it is necessary to be able to select the ratio between carriage travel and mandrel revolutions to extraordinarily high accuracy. The machine must also be capable of holding that accuracy during the entire winding without cumulative error. If this accuracy is not held, gaps or overlaps of filaments may occur, which can lead to weaknesses in the structure.

Another important consideration in choosing a filament winding machine is the type of machine drive required. Four types of machine drives are currently available:

1. Mechanical
2. Hydraulic servo
3. Electric servo
4. Pneumatic

Each of these types of drives has advantages and disadvantages. The majority of machines now in use are of the mechanical type, although many have been modified to use hydraulic motors in order to achieve high thrust at low speeds. Hydraulic and electric servo machines have been used primarily for missile motor construction and for large diameter tanks.

A third major consideration for winding machine selection is the choice of programming controls. The following are the principal types in use:

1. Full mechanical
2. Cam servo
3. Analog servo (variable dial control)
4. Digital servo
5. Computer

The mechanical control is probably the lowest cost and most widely used programming method for filament winding. Its advantages are no cumulative error and good accuracy; its disadvantages are long setup time, lack of flexibility for nonuniform motion, and handling problems when gears and sprockets are dirty.

In recent years the digital servo type of control has become popular [15,16]. By this method the machine maintains a preset program by continually monitoring and adjusting the ratio between carriage drive and mandrel speed.

The type of winding equipment required is, of course, dependent upon the type of product that is being made. McClean discussed the machinery required for eleven types of filament wound structures [17]. The eleven types included hoop or circumferential, helix with a wide ribbon, helix with a narrow ribbon, helix with low winding angle, polar wrap, winding a cone, winding a sphere, winding an ovaloid, and unusual structures. Most structures fall into the category of pipe or tanks that require only helical or circumferential windings. As discussed previously, the optimum angle for pressure pipe is in the order of 55°. Vertical storage tanks (the most common kind) normally are wound at 80°, although some tanks utilize a 90° wrap in combination with mat or spray-up.

Some pipe producers use continuous filament winding equipment, but only a few equipment manufacturers offer continuous filament winding machines [18,19]. One of these is the Drostholm machine manufactured in Denmark. Figure 3.11-5 shows a sketch of this continuous winding machine. In essence, the mandrel consists of a continuous ribbon of steel wrapped onto a rotating framework; the steel ribbon is pushed forward by a cam so that the entire ribbon advances one ribbon width for each rotation of the mandrel. At the unsupported end the ribbon is guided back through the axis of the machine to the beginning.

This endless ribbon gives the effect of a continuously moving surface rotating in front of the stationary filament delivery equipment. In practice, the Drostholm machine is frequently used with chopped glass and resin, along with filament winding bands, to achieve a hoop wound, large diameter tube. The minimum diameter is about 45 cm (18 in). The maximum diameter is 3.6 m (12 ft).

A different type of continuous winding equipment has been used by A. B. Chance Co., who for many years has manufactured electrical utility truck booms and other "hotline" equipment items [20]. This process utilizes a "pultrusion" type system with zero degree reinforcements formed around an internal shape. Circumferential dry windings are then wrapped around the pultruded elements with wet-out of the circumferentially wound fibers being accomplished on the mandrel from the excess resin of the 0° reinforcement.

In addition, specialized, custom-manufactured equipment has been designed and produced for various applications. In the manufacture of high voltage fuse tubes, for example, some manufacturers use multiple winding heads to produce two or three parts at one winding cycle, which reduces the winding time per part. Other specialized machines have utilized automatic mandrel insertion and removal for pipe production.

The introduction of computer controlled multiaxis machines has brought a new dimension to the filament winding machine process and can allow the winding of very odd shapes such as pipe tees or crosses [21]. In the past these shapes could be wound only with specialized equipment. The use of this type of equipment is expected to increase in the future because of the great versatility of digitally programmed machines [22].

Table 3.11-1 shows a list of some filament winding equipment manufacturers with a brief description of the types of equipment available. McClean-Anderson (Milwaukee, Wisconsin) and En-Tec (Salt Lake City, Utah) are two of the largest in this field. These companies have pioneered the shift to computer controls. Figure 3.11-6 shows a photograph of the En-Tec Model 500 Winder.

Accessory equipment such as resin impregnation baths, strand guiding equipment, mandrel insertion and removal units, and curing ovens are sometimes supplied by the filament winding machine manufacturer and are often customized for the particular fiberglass product line by its suppliers.

In Figure 3.11-7, the thickness of a filament wound layer is graphically related to yield, ends/inch, glass

FIGURE 3.11-6. En-Tec computer controlled filament winder.

content and resin specific gravity. By following the vertical "example line," the following can be seen:

1. An increase in glass content from 60% to 70% results in a thickness decrease of about 21%—1.00 to 0.77 mm (0.039 to 0.0305 in) per layer.
2. For an equivalent quantity of glass, the number of ends can vary from 6 to 18 for yields of 225 to 675, respectively.

By following the horizontal "example line," it is shown that, for a given desired thickness per layer, 1.00 mm (0.04 in), the number of ends of 675 yield per inch can vary from 15 to 19 over the glass content range from 50% to 60%.

A resin with higher density will result in a higher composite density if the glass content is held constant. Thus, an increase in resin density will result in a decrease in thickness for a given glass content. In Figure 3.11-7 the vertical example line shows that, for 60% glass by weight, a change in resin density from 1.20 to 1.30 g/cm³ results in a 5% decrease in thickness—1.00 to 0.94 mm (0.039 to 0.037 in) per layer.

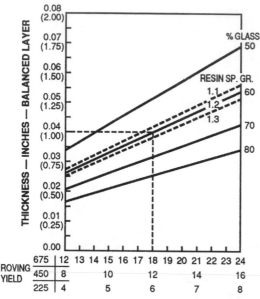

FIGURE 3.11-7. Thickness of filament winding (in inches with values in parentheses being meters × 10³).

3.11.2 Open End Structures

Small Diameter Pressure Pipe

The production of small diameter pressure pipe [below 1.2 m (48 in)] began in the late 1950s with the development of epoxy filament wound pipe for gathering lines for crude oil. A. O. Smith-Inland Corp. (then called Dow-Smith) was the pioneering company in this field [23]. Some USA manufacturers of "line" pipe or other small diameter filament wound pressure pipe are

1. Ameron Co.
2. Ciba-Geigy Corp.
3. Fibercast Co.
4. Fiberglass Resources Corp.
5. Fiberglass Systems
6. Koch Industries
7. A. O. Smith-Inland Co.
8. Texas Fiberglass Products

Whereas most of this type of pipe is made from filament wound fiberglass/epoxy, filament wound fiberglass/vinyl ester is competitive. Most high-pressure small diameter pipe uses the helical wound 54.75° wind

angle construction, but there are exceptions. One process uses approximately ⅓ low angle windings (approximately 9°) with ⅔ high angle windings (approximately 85°). Pipe made by this process is considerably stiffer than the conventional 54.75° helically wound pipe.

Most filament wound pressure pipe is made to comply with specifications of the American Petroleum Institute, the American Society for Testing and Materials (ASTM), or the American Waterworks Association. The most widely applicable specifications are those of ASTM. The principal specifications are as follows:

1. ASTM D2310-71 "Standard Classification for Machine Made Reinforced Thermosetting Resin Pipe"
2. ASTM D2517-73 "Standard Specification for Reinforced Epoxy Resin Gas Pressure Pipe and Fittings"
3. ASTM D2996-71 "Standard Specification for Filament Wound Reinforced Thermosetting Resin Pipe"

The basis for these specifications is the data obtained from two ASTM test methods:

1. ASTM D1599 "Standard Test Method for Short Time Rupture Strength of Plastic Pipe, Tubing, and Fittings"
2. ANSI/ASTM D2143 "Standard Test Method for Cyclic Pressure Strength of Reinforced Thermosetting Plastic Pipe"

The cyclic test method involves the development of a pressure-time regression curve based on cyclic performance at 25 cycles per minute. The shape of the curve is much like a metal fatigue curve. Twelve specimens are normally tested with pressure levels chosen such that failures will result over the range of 1,000 to over 1,000,000 cycles. By plotting the data on log-log coordinates (which invariably gives a straight line), the failure pressure at 150 million cycles (10 years) can be obtained by extrapolation. This test method is the one most frequently used for establishing (with a proper safety factor) the working pressure of the pipe.

Large Diameter Pipe

Large diameter filament wound pipe [above 1.2 m (48 in)] is most often associated with underground installation for either gravity flow, such as sewer pipe, or pressure pipe for conveying fluids. There are a number of constructions used for underground pipe depending upon the type of service. Some of the constructions are

1. Solid wall filament wound pipe
2. Sandwich construction with foam or balsa core
3. Reinforced plastic mortar construction
4. Ribbed construction

Reinforced plastic mortar pipe (RPMP), according to the ASTM definition, is a tubular product containing fibrous reinforcements and aggregate embedded in or surrounded by cured thermosetting resin [24]. The type of aggregate is not specified in ASTM Standards. The usual range is 20–40% by weight of sand aggregate ranging in size from about 20 mesh to 100 mesh.

A new pipe standard (issued in 1981) was drafted by the American Waterworks Association as AWWA C950-81 [25]. This standard has very stringent requirements for underground pipe, either pressure or gravity flow. The standard includes requirements for pressure, for pipe deflection due to overburden, and for buckling. The purpose of the standard is to provide direction and guidance in the selection and purchase of glass-fiber reinforced thermosetting resin pipe and reinforced plastic mortar pipe for use as pressure pipe in water distribution and transmission systems.

Most underground pipe does not use the 54.75° wind angle, even though the pipe may be subject to internal pressure. The resistance of the soil generally prevents

the pipe from expanding in the axial direction, and as a rule, thrust blocks are used to help restrain the pipe at bends. Most of the solid wall pipe is made from a combination of chopped glass fibers and continuous hoop fibers, although aggregate may be added as in the case of reinforced plastic mortar pipe.

Small Diameter Tubing

There are a number of applications (primarily in the electrical industry) for small diameter filament wound tubing ranging from about 6.4 mm (¼ in) up to approximately 150 mm (6 in) diameter. One of the principal uses for this type of tubing is in the manufacture of high voltage fuse tubes. When the fuse is actuated (due to an overload or a lightning bolt), the estimated pressure inside the tube approaches 20 MPa (3,000 psi). Since the tube is open-ended, there is little or no axial load. The usual wind angle is about 70°.

Other applications include high voltage circuit breakers, lighting poles, and small diameter tubing for double insulated hand tools. The circuit breakers often require wound-in threaded metal inserts. Lighting poles are usually tapered tubes. Some manufacturers make a constant wall thickness part that is somewhat more complicated to wind; other producers use normal, constant-pitch winding that causes a greater wall thickness at the smaller diameter, upper end.

3.11.3 Closed End Structures

Vertical Storage Tanks

Filament wound vertical storage tanks (which may be open or closed at the top) are normally produced with an inner liner of chopped glass (spray-up or chopped strand mat) and a surface veil. The veil gives a thin, resin-rich layer [approximately 0.25 mm (0.010 in) thick] containing only about 5% of reinforcement. C-glass surface mat or an unwoven polyester mat are most commonly used.

The purpose of the inner liner is to provide corrosion protection for the filament wound structural layer. Experience has shown that corroding solutions can penetrate through a high glass content laminate, such as a filament wound structure, more rapidly than through a low glass content structure, such as a chopped glass laminate [26].

Table 3.11-1. Some filament winding machine manufacturers.

I. G. Brenner Co. Newark, Ohio 43055	Custom equipment for pipe or tubing (mechanical)
Drostholm Products Vedbaek, Denmark	Continuous winders for pipe or tanks (mechanical). Diameter range: 0.45–3.6 m (18 in–144 in)
Durawound, Inc. Washougla, Washington 98671	Pipe winding equipment (mechanical). Diameter range: below 30 cm (12 in)
Engineering Technology, Inc. Salt Lake City, Utah 84115	Pipe and tank winding equipment (mechanical and electronic/servo) Diameter range: 0–3.6 m (0–144 in)
McClean-Anderson Div. Milwaukee, Wisconsin 53224	Pipe and tank winding equipment (mechanical and electronic/servo) Diameter range: 0–3.6 m (0–144 in)
Snow Machine and Welding Prichard, Alabama 36610	Pipe and tank winding equipment (electronic/servo) Diameter range: 0–3.6 m (0–144 in)

A vertical tank has mainly hoop stresses. Experiments and experience have shown that a measure of axial strength is desirable and that the allowable hoop strain should be less than 0.1% [27]. These conditions can be attained by winding bands at 80° or using layers of chopped glass laminate interspersed with 90° windings. An ASTM standard covering filament wound chemical storage tanks specifies that the corrosion barrier liner construction use the 80° winding angle [14].

In many instances—for example, when the tank bottom must be elevated and cannot rest on a flat surface—a dished or conical head may be used. Ordinarily, the heads are made by hand lay-up with layers of woven rovings and chopped glass, whereas the cylindrical portion of the tank may be filament wound.

In most instances large diameter tanks are wound on a collapsible mandrel that is constructed as a steel cylinder with the wall split along a line parallel to the rotating axis. The opposite side may either be hinged or, for very large cylinders, use the elasticity of the steel wall as a hinge. Movement is usually accomplished by turnbuckles or hydraulic cylinders.

One design uses a cantilever-mounted, tapered cylinder mandrel and a hydraulically actuated push ring to force the filament wound cylinder off the mandrel. This approach is less positive for removal than the collapsible mandrel, since the parts sometimes stick with greater force than the push ring can exert.

Vertical storage tanks are also made from cylinders wound by the continuous Drostholm machine; in this instance the 80° angle is not feasible, and combinations of chopped glass and hoop windings must be used.

Horizontal Storage Tanks and Pressure Vessels

When two closed structural ends are required in a cylinder, such as for a filament wound pressure vessel, the usual procedures are as follows:

1. Wind over a soluble or meltable mandrel
2. Wind over a liner that remains in place
3. Wind two half tanks and join the two halves

During the early period of rocket motor case development, a number of removable mandrel concepts were investigated [17]:

1. Sand filler with polyvinyl alcohol (PVA) binder
2. Meltable wax
3. Water soluble plaster
4. Collapsible, segmented metal

The most practical system was sand filler with PVA binder, which was used for several years for solid fuel rocket motor cases. Currently, the more commonly used technique is a break-out plaster backed by a steel frame.

In applications where diffusion of gases through the laminate must be minimized (such as compressed gas cylinders for a fireman's breathing system or for cryogenic pressure vessels), an impervious liner must be used [28–30]. Ordinarily, an aluminum alloy is used that will strain inelastically (yield) during the first pressure cycle. The aluminum then goes into compression upon release of the pressure.

Metallic pressure vessel design, fabrication, and test procedures normally follow the ASME Section VIII Pressure Vessel Code. ASME Section X covers the design and test procedures for composite plastic vessels. This code requires a destructive pressure test to prove the design. This test is a severe economic penalty for one of a kind or low serial production vessels. As a result there are very few large pressure vessels made from filament wound fiberglass construction despite the light weight, corrosion resistance, and cost advantages compared to stainless steel or other alloys.

The construction of filament wound pressure vessels normally follows the "balanced in-plane" contour system [31]. The isotensoid vessel, using geodesic winding paths, was found equal in performance, but slippage of the windings often occurred in the transition zone between the head and the cylindrical sections.

3.11.4 Miscellaneous Structures

There are other structures made by filament winding that are not normally classified as cylindrical. Spring elements have been made not only by unidirectional moldings but also by filament winding [32]. Coil springs have not been so practical as leaf springs since the coil spring develops torsional shear stresses that unidirectional composites do not effectively resist.

Other filament wound structures and materials that have been investigated include

1. Aircraft tires wound over a washout mandrel [33]
2. Box beam sections for aircraft wings [34]
3. Sheet molding compound (SMC)

Aircraft tires did not prove to be economical. Fixed winged aircraft have not been made by filament winding; however, helicopter blades are predominantly made from filament wound fiberglass [35].

The filament wound sheet molding compound (SMC) was first developed in the early 1960s [36] but did not become commercialized at that time. The product was revived in 1976, primarily for molding structural automotive parts [37].

The molding compound is formed by helically winding continuous strands onto a cylindrical mandrel to a predetermined thickness. Normally a 45° winding angle is used; however, any angle or combination is possible.

After winding, the laminate is slit along a line parallel to the rotating axis, and the sheet material is peeled off the mandrel. This sheet material can be molded in conventional matched dies with standard techniques. The compound will not flow in the direction parallel to the fibers; hence, the charge must be cut to nearly fill the mold cavity.

Either polyesters or epoxies can be used in making the filament wound SMC. Normally, the epoxy is advanced to a non-tacky state of cure (B-staged). Polyesters are thickened with chemical agents as well as standard chopped fiberglass SMC.

3.11.5 References

1. Anonymous. "Polaris A-3 Fiber Glass Case Being Developed Rapidly," *Missiles and Rockets*, p. 36 (January 29, 1962).

2. SHIBLEY, A. M., H. L. Perritt and M. Eig. "A Survey of Filament Winding: Design Criteria and Military Applications," *OTS Plastec Report No. 10*, 251 pp. (May 1962).

3. Anonymous. "Filament Winding Goes Commercial," *Machine Design*, 36:130 (May 7, 1964).

4. ROLSTON, J. A. "Fiberglass Composite Materials and Fabrication Processes," *Chemical Engineering*, 87:96 (January 28, 1980).

5. MARTIN, J. D. and C. S. Richter. "A Marketing Approach to the Development of the RP/C Lighting Pole Market," *SPI 29th Annual Conf.*, Section 3-A, 6 pp. (February 1974).

6. FINK, N. H. "Application of Fiber Glass Reinforced Plastics in Transmission and Distribution of Electrical Power," *SPI 22nd Annual Conf.*, Section 15-B, 10 pp. (February 1967).

7. WEINGART, O. "Low Cost Composite Blades for Large Wind Turbines," *SPI 35th Annual Conf.*, Section 17-A, 4 pp. (1980).

8. YOUNG, R. E. "Pressure Vessels from Plastic Bonded Glass Fibers," *Sagamore Conf.*, p. 370, OTS PB 161443 (August 1959).

9. YOUNG, R. E. "History and Potential of Filament Winding," *SPI 13th Annual Conf.*, Section 15-C, 6 pp. (1958).

10. SONNEBORN, R. H. and A. G. H. Dietz. *Fiberglass Reinforced Plastics*. Reinhold Publishing Corp., New York, NY (1954).

11. YEAGER, R. W. and J. R. Sullivan. "Design Curves for Filament Wound Rocket Motor Cases," *SPE 20th Annual Conf.*, Section XIX-1, 17 pp. (January 1964).

12. HOFEDITZ, J. T. "Structural Design Considerations for Fiber Glass Pressure Vessels," *Modern Plastics*, 41:127, 142 (April 1964).

13. PARADY, V. "How to Reinforce Holes in Filament Wound Structures," *Materials in Design Engrng.*, 61:108 (February 1965).

14. ASTM D3299-74, "Standard Specification for Filament Wound Glass Fiber Reinforced Polyester Chemical Resistant Tanks," *American Society for Testing and Materials*, Philadelphia, Part 36.

15. WESCH, L. "Recent Developments in Filament Winding Equipment," *Plastics and Polymers*, 37:149 (April 1968).

16. KIRKLAND, C. "Computer Controlled Filament Winding—Why It's Catching on Fast," *Plastics Technology*, 26:77 (May 1980).

17. McCLEAN, W. G. "How to Select a Filament Winding Machine," *Plastics Design and Processing*, 5:18 (January 1965).

18. POULSEN, U. "Production Plant for Large-Scale Manufacture of Cylindrical Tanks," *Plastics and Polymers*, 36:143 (April 1968).

19. Anonymous. "Filament Wound Pultrusions Bring New Performance Levels to Pipe," *Modern Plastics*, 53:102 (April 1976).

20. FINK, M. H. "Continuous Manufacture of Large Reinforced Plastics Hollow Shapes," *SPE Regional Tech. Conf.*, Cleveland Section (September 1969).

21. ZEIER, D. J. "Filament Winding Potential with Advanced Multi-Axis Computer Controlled Filament Winding Machines," *Proc. from the Second Conference on Advanced Composites, November 18–20, 1986, Dearborn, MI*, ASM International, p. 33 (1986).

22. WELLS, G. M. and K. F. McAnulty. "Computer Aided Filament Winding Using Non-Geodesic Trajectories," *Proc. Sixth Int. Conf. on Composite Materials (ICCM & ECCM)*, F. L. Matthews et al., eds., Elsevier Applied Science, London, p. 1.161 (1987).

23. PFLEDERER, F. R. "Cyclic Pressure Testing of Reinforced Plastic Pipes," *SPI 17th Annual Conf.*, Section 1-D, 6 pp. (February 1962).

24. ASTM D3262-76, "Standard Specification for Reinforced Plastic Mortar Sewer Pipe," *American Society for Testing and Materials*, Philadelphia, Part 34.

25. AWWA C950-81, "AWWA Standard for Glass Fiber Reinforced Thermosetting Resin Pressure Pipe," *American Water Works Assn.*, Denver, Colorado.

26. ATKINSON, H. B. "Proposed Product Standard for Filament Wound FRP Chemical Resistant Tanks," *SPI 24th Annual Conf.*, Section 4-B, 10 pp. (February 1969).

27. ISHAM, A. B. "Design of Fiberglas Reinforced Plastic

Chemical Storage Tanks," *SPI 21st Annual Conf.*, Section 16-E, 8 pp. (February 1966).

28. Anonymous. "Fiberglass Filament Wound Reinforced Composites Approved by COT for Compressed Gas Cylinders," *Materials and Energy for Industry (M.E.I.)*, p. 4 (January 1977).

29. JONES, R. H. and A. Kauffman. "Filament Overwrapped Metallic Cylindrical Pressure Vessels," *AIAA/ASME 7th Structures and Materials Conf.*, p. 52 (April 1966).

30. McLOUGHLIN, J. R. "Low Temperature Properties of High Performance Composites," *SPI 30th Annual Conf.*, Section 18-D, 6 pp. (February 1975).

31. DARMS, F. J., R. Molho and B. E. Chester. "Improved Filament Wound Construction for Cylindrical Pressure Vessels: Vol. I—Structural Analysis and Materials and Processes," *OTS AD 436272*, 201 pp. (March 1964).

32. WARNER, G. G. and A. Torosian. "Glass/Epoxy Spring Is 80% Lighter Than Steel," *Plastics Design Forum*, 5:14 (July/August 1980).

33. SALCEDO, F. W. and T. J. Reinhart. "Development of Filament Wound Aircraft Tires," *12th Nat'l. SAMPE Symposium*, Section AS-3, p. 24 (October 1967).

34. TANIS, C. "A New Manufacturing Method for Airframe Structures," *12th Nat'l. SAMPE Symposium*, Section AS-6, 32 pp. (October 1967).

35. LOUD, S. N., JR. "Advanced Composites Applications of S-2 Glass Fiber," *SPI 34th Annual Conf.*, Section 20-F, 8 pp. (February 1979).

36. IRVING, R. R. "Filament Winding Goes after Commercial Markets," *The Iron Age*, p. 159 (November 14, 1963).

37. ACKLEY, R. H. "XMC Structural FGRP for Matched Metal Die Molding," *SPI 31st Annual Conf.*, Section 16-C, 4 pp. (February 1976).

SECTION 3.12

Pultrusion

Introduction

The origin of the word *pultrusion* is obscure, but the meaning is quite plain. Thermoplastic or aluminum extrusions are produced by pushing material through a die. The same shapes can be made with glass reinforced plastics, but the glass fibers (first wet with a liquid resin) must be pulled through the die, hence, the name—abbreviated as "pul-trusion."

The pultrusion process, which began around 1948, was first patented in 1951 [1]. The process requires continuous fiber reinforcements and low viscosity (normally liquid thermosetting) resins. Fiberglass rovings are predominant, although graphite fibers [2] and Kevlar [3] have been used in developmental quantities or in blends with glass fibers. The number of rovings used may range from one to a very large number. The greatest number known is 5,000 rovings in one machine, making a 42.5 cm (17 in) square box beam with 2.54 cm (1 in) thick walls.

Pultrusion is a truly continuous process: with an automated cut-off saw, a pultrusion line can run with virtually no attention except for occasional checking for glass breakouts and resin levels.

The first products made by pultrusion were fishing rods, which are still a major business. The high tensile strength of pultruded rods combined with other properties, such as electrical insulation, corrosion resistance, and light weight, have led to a number of other applications: electrical motor or generator insulating spacers, tool handles for working on high voltage equipment, structural shapes (angles or channels), subway third rail covers, CB antennas, and a large number of other uses. Figure 3.12-1 shows a pultruded demister blade (such as used in air pollution scrubbers) emerging from the puller.

3.12.1 Equipment

The basic parts of the pultrusion process are

1. Reinforcement handling
2. Resin impregnation
3. Pre-die forming
4. Shaping and cure of the resin
5. Pulling and cut-off

In the early 1960s the hot die process was developed. This process uses heated dies shaped to the desired cross section (such as an angle or channel) which form the composite while it is cured. This improvement enabled the use of mat or cloth additions to the pultruded section for greater transverse strength. Very intricate cross sections have been developed, i.e., I-beams with sizes up to 30 cm (12 in) or larger. A photograph of a pultrusion line with hot die curing systems is shown in Figure 3.12-2.

Reinforcements Handling Equipment

Bookshelf creels are the usual style with roving package yields of 540, 270, or 135 meters per kilogram (225, 113, or 56 yards per pound). Most producers pull the glass from the inside of the package rather than unrolling the package and pulling from the outside. The inside pulling technique gives a slight twist (one turn per inside circumference), but the effect is minimal. Since the package is stationary, the outside roving end can be tied to the inside end of a new package to obtain automatic transfer. A typical creel arrangement is shown in Figure 3.12-3. The fiberglass strand guides are usually ceramic or ceramic coated metal for smoothness and wear resistance.

Continuous strand mats of 0.50 kg/m² (1½ oz/ft²) together with rovings are the normal reinforcements used for flat sections or thin-walled shapes. Most fabricators sandwich a layer of rovings between each two layers of mat. The mat is not strong enough to pass through the wet-out bath without the added strength of the rovings. Mat tensile strength has been a serious problem in the past but is somewhat less so at this time.

FIGURE 3.12-1. Pultruded demister blade emerging from puller.

FIGURE 3.12-2. Pultrusion line with hot die curing.

Mats must be unrolled, and the end of one roll must be fastened together (usually hand stitched) with the new roll on the fly. A roll of mat is normally about 90 m (295 ft) long. This roll will last about 1 to 3 hours, depending on the processing speed. If many mat layers are used, the end-joining frequency can become burdensome.

Mats must be handled and guided more gently than rovings. Sharp turns are not feasible, and the tension must be kept to a minimum. The minimum width is about 10 cm (4 in). When transverse strength is needed on narrower sections, spun rovings can be used. This is a special roving with numerous loops that add bulk; it is packaged similarly to standard roving.

Resin Impregnation

Pultrusion resins are predominantly unsaturated polyesters, but epoxies and silicones have been used in a few special purpose applications [4]. Recently, thermoplastic resin pultrusion has been examined [5]. The range of polyester types is from lowest cost orthophthalic to premium grades, such as bisphenol A or vinyl ester. The most common types of filler, which range from none to about 20%, are calcium carbonate, alumina trihydrate, clay, and antimony trioxide (with halogenated fire retardant resins). The upper limit is controlled by resin viscosity; at high filler loadings the rovings will not wet adequately. Pigments are often

FIGURE 3.12-3. Pultrusion creel.

FIGURE 3.12-4. Hot die cure in pultrusion.

added to the resin for coloration and ultraviolet resistance. An internal mold release is always required since the die is continuously wiped by the moving composite. Mold releases include organic phosphates, lecithin, and metallic stearates.

Resin impregnation baths are usually 0.9 to 1.8 m (3 to 6 ft) long and as wide as necessary. [Filament winding resin baths are normally much shorter, 30 to 45 cm (12 to 18 in), even though the linear speed through the bath may be fifty to one hundred times faster. The probable reason for this disparity is that in filament winding much of the wet-out occurs on the mandrel.] In pultrusion, as soon as the rovings reach the die cavity, the remaining air is trapped in the laminate.

Most resin baths have bars between which the rovings (and mats) pass over and under. If multiple layers are required, the pairs of roving and mat layers should be kept separate through the impregnation bath and combined thereafter. Cloth layers (such as woven rovings) can be substituted for rovings to enable higher transverse strength than with continuous strand mat.

Surface mats or printed overlays are sometimes added for surface appearance or to improve corrosion resistance. Glass-fiber surface mats (C veil) have no strength in the wet state; they should be fed onto the laminate at the entrance to the curing die. Polyester surface mats (unwoven cloth), which have a high wet strength, can be drawn through the impregnation bath.

However, polyester surface mats are more expensive than fiberglass.

Pre-Die Forming

During the forming process in pultrusion, a sharp change of direction in one operation is difficult, if not impossible; gradual changes from a flat shape to a U-shaped or V-shaped cross section should be accomplished through several forming steps. The pre-die formers need not be elaborate; bent wires or slotted plates are the most common pre-die formers.

Shaping and Curing

The hot die for curing (see Figure 3.12-4) is one of the most costly parts of the process [6]. The dies—normally 0.9–1.5 m (3–5 ft) long—are made from alloy steel, heat treated to a high hardness, and polished to as smooth a surface as possible. Specifications for many pultruders call for an RMS surface profile of 0.1 to 0.2 μm (4 to 8 microinches) or less. After polishing the surface, a layer of hard chrome plate should be applied to a depth of at least 0.05 mm (2 mils). It is advantageous to design the tool for end-for-end reversal, since the front end suffers the most wear. After the chrome plating has worn off, the tool can be turned around, thus doubling the useful life. As a general rule, a good die will run about 30 km (100,000 ft) of material before stripping and rechroming are necessary.

The surface smoothness of the die is critical, especially in the zone where the resin has reached a gel state but has not become fully cured. If resin sticks to a scratch on the die in that area, it will scrape a roughened section on the part, and broken particles of cured resin will emerge from the die in that portion of the structure. The remedy is to stop the pultruder for about a minute to allow the resin to cure and bond to the resin spot stuck to the die. Often this technique will clear the die so that a smooth surface on the cured part is again restored. Speeding up or slowing down the operation may move the sensitive gelled area away from the scratch that was causing the problem.

Dies are normally split on a horizontal plane parallel to the pulling direction, although one-piece dies are often used for round rods. These dies are made by gun drilling and honing. They can even be chrome plated by special techniques. The advantage of a one-piece die is

the improved smoothness resulting from no parting line in the mold.

A comparison of several types of dies and cure systems is presented by Meyer [7]. This report lists advantages, disadvantages, and comments for various types of pultrusion processes including variables such as

(a) Pulling—intermittent or continuous
(b) Dies—rigid metal, compression molding, Teflon film wrap, liquid metal
(c) Cure—hot die, oven, radio frequency

Pulling and Cut-Off

Most pullers and cut-off saws are custom designed and fabricated but may be purchased as individual items. Two types of pullers are used in pultrusion. The most common type grips the pultruded part between two rubber-cleated chain drives with the gripping pressure between the chain drives maintained either mechanically or hydraulically by pressure cylinders. The usual chain length is 1.2 to 1.8 m (4 to 6 ft). The other type of puller is an intermittent back-and-forth vise clamp which grips on the forward stroke and releases on the back stroke. Two of the intermittent clamps may be combined to give a continuous motion, although some pultruders prefer the intermittent action. An advantage of the reciprocating clamp over the rubber-cleated caterpiller-type drive is its ability to handle thin-walled complex shapes with under-cuts, since the clamp slides over the part during the upstream stroke. A metal support block can be used inside the part with the support block moving back and forth with the puller.

The "flying" cut-off saw uses either wet or dry silicon carbide grit or a diamond blade. The pultruded section must be clamped to the saw table during cutting so that the cutting blade and table will then move with the part, hence, the name *flying*. In most shops the part length is controlled by an adjustable microswitch that initiates the saw action automatically.

For some types of structures, the thermal shrinkage causes warping of the part after the cured section emerges from the die. Concave sections, such as channels or U-sections, tend to close up during cooling. To prevent the shrinkage, the part may be held in place with wooden or metal blocks. Faster cooling may be obtained with water sprays placed between the hot die and the puller.

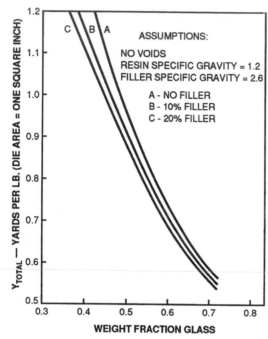

FIGURE 3.12-5. Glass-fiber roving yield for pultrusion.

3.12.2 Uniaxial Reinforced Structures

Roving Yield Calculations

Relationships of glass, resin, filler, and stripping die area are provided in reference [8]. The density of a filled resin may be calculated from the relation

$$\varrho_m = \frac{W_f}{\varrho_f} + (1 - W_f)/\varrho_r(1 + H_m)$$

where

ϱ_m = density of resin and filler mixture
W_f = weight fraction of filler
ϱ_f = intrinsic density of filler
$1 - W_f$ = weight fraction of resin
ϱ_r = density of resin
H_m = void fraction of resin plus filler

If the void content is unknown (as is often the case), the density may be determined by experiment from a cured casting of the mixed resin and filler combination. The equation is

$$\varrho_m = W_m/V_m$$

where

W_m = measured weight of casting
V_m = measured volume of casting

From the measured density the void fraction can be calculated from

$$H_m = \frac{1}{\varrho_m[W_f/\varrho_f + (1 - W_f)/\varrho_r]} - 1$$

In a similar fashion the volume proportion of glass fibers can be calculated from the equation for volumetric conversion (using the resin/filler mixture density):

$$V_g = \frac{W_g/\varrho_g}{[W_g/\varrho_g + (1 - W_g)/\varrho_m](1 + H_c)}$$

where

V_g = volume fraction of glass fibers
ϱ_g = intrinsic glass density
W_g = weight fraction of glass fiber
H_c = void fraction of the composite (glass and filled resin)

The curing die area can then be determined from

$$A = \frac{1 + W_g(\varrho_m/\varrho_g - 1)}{36W_g\varrho_m Y_{tot}}$$

where

A = curing die area in square inches
Y_{tot} = total roving yield in yards/pound ($\times$ 0.2056 m/n) = Y_g/N

where

Y_g = roving yield of individual packages
N = number of packages

Figure 3.12-5 shows the plot of Y_{tot} against the glass weight fraction for several ratios of resin (density =

1.2 g/cm³) and clay filler. The die size is assumed to be 1 sq in (6.45 × 10⁻⁴ m² = 6.45 cm²).

The use of the preceding equations can be seen from the following illustration.

Example of Calculation of Roving Strands
Assumptions:

glass content = 65%
filler and void content = 10% clay with 10% voids
die area = 0.75 sq in (4.83 × 10⁻⁴ m² = 4.83 cm²)
glass density = 0.0920 lbs per cu in (2.5465 × 10³ kg/m³)
resin density = 0.0433 lbs per cu in (1.1985 × 10³)
filler density = 0.0942 lbs per cu in (2.6074 × 10³)

From the equation for resin/filler density (ϱ_m) we have

$$\varrho_m = \frac{1}{0.10/(0.00942) + (1 - 0.10)/(0.0433)(1 + 0.10)}$$

$$= 0.0416 \text{ lbs per cu in } (1.1515 \times 10^3 \text{ kg/m})$$

The total yield for a 1 sq in die is

$$Y_{tot} = \frac{1 + 0.65(0.0415)/(0.092) - 1}{(36)(0.65)(0.0415)(1)}$$

0.663 yd per pound (1 sq in die)
234 m per Newton (1 sq m die)
0.0234 m per Newton (1 sq cm die)

For a smaller die, the yield will be greater:

$$Y_{tot} = \frac{0.633}{0.75} = 0.884 \text{ (0.75 sq in die)}$$

$$= \frac{0.0234}{0.75} = 0.0312 \text{ (0.75 sq cm die)}$$

Assuming the use of roving with a yield of 225 yd per pound, the number of packages required is

$$N = 225/0.663 \text{ (1 sq in die)}$$

$$= 254 \text{ packages}$$

Uniaxial reinforced structures are limited to those where little or no transverse strength is required. Fish-ing rods, electrical rods and spacers, and similar items are typical examples. Very small diameter [0.75 mm (0.030 in)] rods were developed for high altitude balloon tethers in the mid-1960s. These were made from S-glass reinforced epoxy resin and were cured by passing the wet strand through a molten lead alloy bath at a pulling speed of about 3 m/min (10 ft/min). Larger diameter rods [12.7 to 19 mm (½–¾ in)] have been used for military radar antenna guy wires [9].

3.12.3 Biaxially Reinforced Structures

Laws of Combined Action

Pultrusion may be made with parallel fibers in the principal stress direction and mats, cloth, or filament winding in the minor stress direction. This is the usual technique for structural shapes such as angles, channels, or I-beams. The effect of the mat layers on strength or stiffness may be calculated based on the volume percentage of mat and rovings in the laminate.

In practical structures made from pultruded elements, the tensile or bending strengths are seldom realized due to failures at joints or from interlaminar shear failures. The problems of joints (both strength and high cost) have had inhibiting effects on the wider use of structures fabricated from pultruded shapes. Combinations of pultrusion and filament winding have been used for a number of years [10], but the technology is not widespread.

3.12.4 References

1. HOWALD, A. M. and L. S. Meyer. "Shaft for Fishing Rods," U.S. Patent No. 2,571,717 (assigned to Libby-Owens-Ford) (October 16, 1951).
2. HILL, J. E., J. C. Goan and R. Prescott. "Properties of Pultruded Composites Containing High Modulus Graphite Fibers," *SAMPE Quarterly*, 4:21–27 (June 1973).
3. Anonymous. "Pultrusions Give Designers Plastic Alternative to Metal Extrusions," *Product Engineering*, 48:34–35 (December 1977).
4. KRUTCHKOFF, L. "Pultrusion, Part 2–Desirable Resin Properties Are High Heat Distortion, Fast Cure, Good Wet-Out," *Plastics Design and Processing*, 20:37–41 (August 1980).
5. NOLET, S. C. and J. P. Fanucci. "Pultrusion and Pull-Forming of Advanced Composites," *Proceedings from a*

Workshop on Design, Manufacture and Quality Assurance of Low Cost, Lightweight Advanced Composite Molds, Tools and Structures, Florida Atlantic University, Boca Raton, FL (October 17–19, 1988).

6. TICKLE, J. "Lower Cost Tooling Carves Out a Big Market for Pultrusion," *Plastics Engineering*, 31:36–38 (October 1975).

7. MEYER, L. S. "Pultrusion," *SPI 25th Annual Conf.*, Section 6-A, 8 pp. (February 1970).

8. ROLSTON, J. A. "Process and Economic Factors for Pultrusion," *SPI 33rd Annual Conf.*, Section 8-G, 5 pp. (1978).

9. HALSEY, N., R. A. Mitchell, D. E. Marlowe and L. Mordfin. "Non-Metallic Antenna-Support Materials. Pultruded Rods for Antenna Guys, Catenaries and Communications Structures," AFML TR 76-42 (May 1976).

10. FINK, M. H. "Continuous Manufacture of Large Reinforced Plastics Hollow Shapes," *SPE Regional Tech. Conf.* (Cleveland Section) (September 1969).

SECTION 3.13

Rheology of Fiber Filled Polymers

3.13 RHEOLOGY OF FIBER FILLED POLYMERS

A. B. METZNER

3.13.1 Introduction

In this section the flow behavior of concentrated suspensions that may be of interest in processing composites and similar polymeric materials is discussed. Succinct synopses are arranged at the end of each subsection.

In general, the state of the art is a good one: a small number of experiments on a new formulation frequently suffices for a general prediction of flow behavior when solids are suspended in viscous molten polymers, and, in many instances, even *a priori* predictions of the viscosity are possible. Less viscous systems are much more difficult; several remaining problems and research areas are identified.

We shall begin by considering the simplest possible systems and then proceed to progressively more complex behavior. The reader who is interested in only some specific later subsections is nevertheless encouraged to browse from the beginning, in order to develop a general appreciation for the quality of the work on this subject and for its limitations.

At the outset we should clearly define the range of systems we are considering. We shall consider all kinds of suspending fluids, i.e., viscoelastic polymer melts as well as Newtonian liquids. We shall consider both shearing and elongational modes of deformation of the system and low, as well as finite, deformation rates. We shall consider fibrous as well as spherical particles and the effects of both wide and narrow particle size distributions. We shall consider, as particles, only solid particles; emulsions, foams, and suspensions of sticky, solvated particles or of deformable drops will not be treated.

3.13.2 Suspensions in Viscous Liquids

If the fluid phase surrounding the particles is of very high viscosity, then viscous forces imposed by the fluid on the particles when a flow process is initiated will be so great that particle-particle interactions can only be negligibly small. Consequently, no long-range "structures" can build as a result of the forces between the particles, and they act as truly inert suspended species. This lack of structure leads, then, to the simplest possible suspension behavior. It is of considerable importance in its own right, as we shall see, and also represents a firm foundation on which to build as more complex fluid-particle systems are subsequently considered.

An immediate question that arises has to do with the minimal fluid viscosity level required to overpower all interparticle forces. A precise answer to this query is not known, but it is seen that suspensions in molten polymers quite generally fulfill this condition: when the fluid phase viscosity exceeds about 10^3 poise, there is usually no evidence of interparticle effects. Conversely, these latter effects frequently dominate when the suspending fluid has a viscosity of the order of a few poise or less. More importantly, we shall be able to provide simple tests for the absence of interparticle forces.

Suspensions of Spherical Particles of Narrow Size Distribution

General references on this subject include the classic works of Einstein [1], the text by Happel and Brenner [2], a substantial theoretical paper by Frankel and Acrivos [3], contributions by Mooney [4], and a superlative review by Thomas [5].

Figure 3.13-1, taken from Thomas [5], depicts the increase in viscosity that is obtained as progressively more solids are added to the suspending liquid. There are several conclusions to be drawn from this work:

1. There appear to be no effects of particle size or fluid viscosity; both parameters were varied substantially, and one sees that the relative viscosity (the ratio of

the viscosity of the suspension to that of the suspending fluid) is independent of these variables. Of course, the viscometric instrument used must have clearances that are large compared to the particle size so that the effects of the reduced particle concentration near any planar wall [6] are negligible or can be accounted for.

2. The *volumetric* concentration level controls the viscosity level, and care should be taken to *express concentration in these units.*

3. The Einstein equation

$$\mu_r = \mu_s/\mu_o = 1 + 2.5\phi \qquad (3.13\text{-}1)$$

in which ϕ denotes the volume fraction solids, may be an accurate representation of the behavior of the suspension over only a vanishingly small range of solids concentrations.

4. As concentration levels corresponding to a dense packing of the solid particles are approached, there

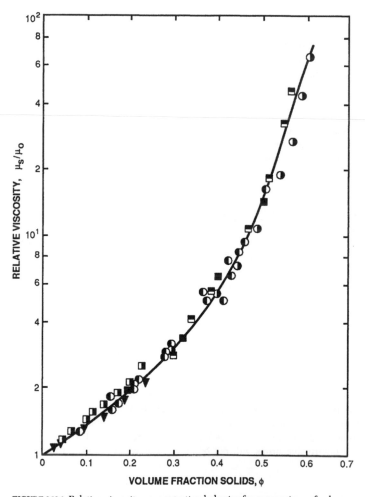

FIGURE 3.13-1. Relative viscosity-concentration behavior for suspensions of spheres having narrow size distributions. Particle diameters range from 0.1 to 440 microns. The line shown depicts Equation (3.13-3). Reproduced from Thomas [5] with permission.

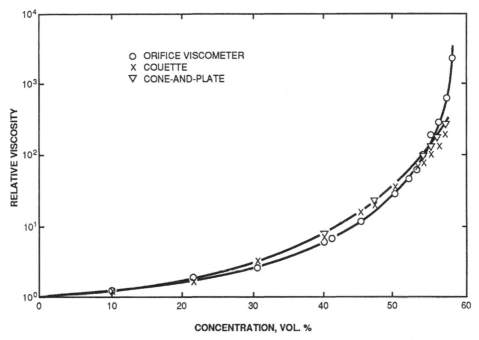

FIGURE 3.13-2. Relative viscosity-concentration curve from Patzold [7] showing results for three types of viscometers. Reproduced with permission.

is no longer sufficient fluid in the system to lubricate the relative motion of particles, and the viscosity rises to infinity, as expected.

Figure 3.13-2, taken from Patzold [7], shows results that are qualitatively similar to those of Figure 3.13-1. Figure 3.13-2 has the virtue of comparing three different instruments: a calibrated orifice flow device, a rotational bob-and-cup (Couette) viscometer, and a rotational cone-and-plate unit. Quantitatively, the results differ markedly from those of Figure 3.13-1, especially at the higher solids concentrations. Patzold [7] notes that, as the particle packing density approaches its maximum possible value, a shearing of the system implies momentary volumetric dilations: a layer of faster moving particles cannot simply slide past its slower moving neighbors, but rather individual particles must rise up and over their neighbors. Brady and Bossis [8] have recently *predicted* the motion of ensembles of particles forming layers that slide over each other in shearing flows. The presence of a rigid wall would, of course, inhibit such motion. Since the ratio of the in-strumental clearance or gap dimension to the particle size was as small as 4 at the center of the cone-and-plate apparatus, it seems likely that such "wall effects" may be responsible for the higher viscosity levels depicted in Figure 3.13-2. This comment is not made derisively but to illustrate the experimental difficulty of obtaining definitive results.

Such a sensitivity to experimental effects suggests that the single curves of Figures 3.13-1 and 3.13-2 should probably be replaced by a "band" of finite width: Cheng [9] noted that the exact position of one's data in such a band will also depend dramatically upon the concentration uniformity of the system. The exponential rise of the viscosity with solids concentration, at high solids loadings, implies the non-cancellation of even small regional concentration variations. We should also note that the coefficient of thermal expansion of solids and liquids differs markedly and that this effect may lead to much larger variations of viscosity with temperature than expected with homogeneous liquids. This effect could lead to viscosity decreases as well as increases with decreasing temperature.

A substantial variety of equations has been suggested for description of the viscosity concentration behavior of suspensions. The only theoretical equation that appears to depict trends correctly in the region of high concentration levels is that of Frankel and Acrivos [3]:

$$\mu_r = \frac{9}{8} \frac{(\phi/\phi_m)^{1/3}}{1 - (\phi/\phi_m)^{1/3}} \qquad (3.13\text{-}2)$$

where ϕ_m = the volume fraction of solids corresponding to the maximum packing density of the particles.

Thomas [5] recommends the empirical equation

$$\mu_r = 1 + 2.5\phi + 10.05\phi^2 + A \exp(B\phi) \qquad (3.13\text{-}3)$$

with $A = 0.00273$ and $B = 16.6$, which is similar in its predictive behavior to the Mooney equation [4]:

$$\log \mu_r = \frac{5}{2}\left(\frac{\phi}{1 - K\phi}\right) \qquad (3.13\text{-}4)$$

in which the parameter K is determined empirically and usually is in the range $1.35 \le K \le 1.91$.

Remarkably, however, the best empirical expression available is probably the very simple form originated by Maron and Pierce that was carefully evaluated by Kitano, Kataoka, and their co-workers [10,11]:

$$\mu_r = [1 - (\phi/A)]^{-2} \qquad (3.13\text{-}5)$$

in which the single empirical constant A has a value of 0.680 for suspensions of smooth spheres in a liquid.

Figures 3.13-1 and 3.13-2 were selected for inclusion because they represent unusually good experimental measurements. Since even these data differ so appreciably, it really is virtually impossible to compare Equations (3.13-2–3.13-5) rigorously; they all have been used successfully over some restricted concentration range or for some specific suspensions. Rather than being able to make the "scientifically best" choice, one must choose the one that appears to have the greatest potential or has been demonstrated to do the best job of portraying the data of greatest interest. This latter condition pertains to Equation (3.13-5). The work of de Sousa [12] supports Equation (3.13-5) for filled HDPE melts. In the following discussions we will evaluate the utility of Equation (3.13-5) in some detail.

Effects of a Distribution of Particle Sizes

Most systems of interest are, of course, more complex than the specially prepared monodisperse suspensions considered thus far. Polydispersity shall be considered in two steps.

First, let us note the observation by Eveson [13] that small variations in particle diameter seem to be of no consequence; mixtures of spheres having a four-fold variation in particle size were reported to show less than a 6% change in relative viscosity for volume fractions of solids as great as 0.20. Whereas the absence of any effects over the four-fold range studied may seem surprising, this observation does agree with the mechanistic expectation that the relative motion of particles past each other in dilute systems may not be affected greatly by modest changes in their size. Thus, the conclusion is reached that, in dilute suspensions ($\phi < 0.20$), modest changes in particle size are of no consequence.

In highly concentrated suspensions two mechanisms might be expected to lead to a far greater importance of variations in particle size. First, if the system must dilate locally to enable one layer of particles to slide past a layer of slower moving neighbors, then even modest differences in particle size might be expected to affect this relative motion appreciably. Second, major changes in the maximum possible packing density of the particles—ϕ_m in Equation (3.13-2), $1/K$ in Equation (3.13-4), and the term A in Equation (3.13-5)—will occur if a wide distribution of particle sizes is employed. To pick an absurdly simple but relevant example, it is easy to increase the solids loading in an ensemble of closely spaced basketballs by placing golf balls in the interstices and to continue this solids loading process by then placing sand grains in the voids between the golf balls, and so on.

Several things are clear from the preceding qualitative example. First, the size ratio of one particle to the next is an important parameter, and this ratio should be quite large if we wish to achieve very high solids concentration levels. Second, if an infinitely wide range of particle sizes is available, then the total solids loading that can be achieved in this way is enormously greater than the 68% level indicated by Equation (3.13-5). Third, in the practice of polymer processing operations, for example, there are real constraints on the size range that can be employed; therefore, a trimodal particle size distribution may be the best that one could normally employ.

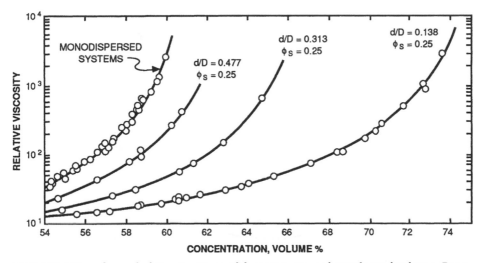

FIGURE 3.13-3. Dependence of relative viscosity on solids concentration and particle size distribution. Reproduced from Chong, Christiansen, and Baer [16] with permission of John Wiley & Sons, New York (copyright owner).

Farris [14] has considered the optimal concentration ratios in some detail. Henderson et al. [15] list the results of theoretical calculations of the effect of particle size modality upon the relative viscosity of suspensions having a solids volume fraction of 0.66 (for spheres) as follows:

Modality	Relative Viscosity
Uni	1,200 [from Equation (3.13-5)]
Bi	51
Tri	30
Tetra	23

This table also shows that, if the purpose of a multimodal distribution is to reduce viscosity rather than to increase solids concentration, the incremental gain beyond a trimodal distribution becomes relatively modest.

Figure 3.13-3 depicts the dramatic effects on viscosity of progressive changes in the particle size ratio in a bimodal suspension of spheres (of diameters denoted by "*d*" and "*D*") and compares the results with the relative viscosity of a unimodal suspension. The ratio, ϕ_s, which denotes the volume fraction of small spheres in the mixture, was kept constant at 25% of the total solids in the system. The data for the unimodal system in

Figure 3.13-3 all fall well above the results shown earlier in Figure 3.13-1 and also well above the predictions of Equations (3.13-1) or (3.13-5) (with $\phi_m = 0.680$). As the reasons for this result are unclear, we cannot be certain that the effects of a bimodal particle size distribution will always be as great as depicted by these data, but even changes in viscosity that were to be far less dramatic might be worth exploiting in order to retain processability of a suspension as the total solids concentration is increased. Figure 3.13-4, from the same reference, shows calculated smooth lines and experimental relative viscosities for bimodal suspensions. Again, the effects depicted are dramatic. It is also impressive to note that the minimal total solids concentrations considered in Figures 3.13-3 and 3.13-4 are 54% and 55% by volume, respectively.

In summary of this discussion, the distribution of particle sizes has very little effect on suspension viscosity when the volumetric loading of solids is below 20%. At high concentration levels the effects are of enormous magnitude, and extraordinarily high solids concentrations can be achieved by using multimodal distributions of particle size. The available results, while impressive, are probably too limited to produce definitive design equations, and experimental data would be needed to design with confidence.

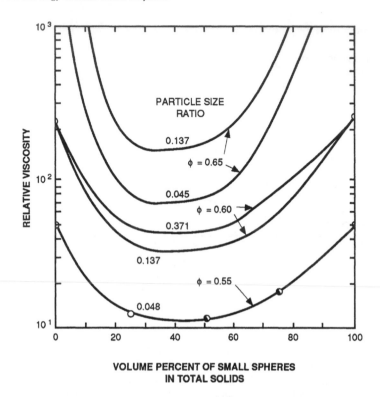

FIGURE 3.13-4. Calculated viscosities and data showing dependence of relative viscosity upon particle size ratio for bimodal suspensions of spheres [16]. Reproduced with permission of the copyright owner, John Wiley & Sons, New York.

Effects of Particle Geometry
(Aspect Ratio < 30)

What happens when the suspended solid particles are not spheres but are cubic or other rough crystalline forms, or when we employ short fibers?

Figure 3.13-5 answers at least a part of this question in an elegant manner. It shows that data for talc, for natural and precipitated calcium carbonate particles, and for a variety of fibers having aspect ratios varying from 6 to 27 inclusive can all be correlated using Equation (3.13-5)—provided the value of the single adjustable parameter A is fitted to the properties of the particular suspension at hand. This correlation is potentially a very powerful generalization and, if confirmed by other data, would suggest that the viscosity of almost any suspension can be predicted over appreciable ranges of concentration on the basis of a deter-

mination of the single parameter A. The range of data included in Figure 3.13-5 is not so great as one might wish, but it does cover much of the range of practical interest. If even a single experimental determination is unavailable, one can approximate A using their results:

Aspect Ratio	Value of A
1.0	0.68 (for smooth spheres) to 0.44 (for rough crystals)
6 to 8	0.44
18	0.32
23	0.26
27	0.18

Figure 3.13-6, also from Kitano et al. [10], shows that, for short-fiber suspensions, the parameter A decreases linearly with aspect ratio. It is important to note

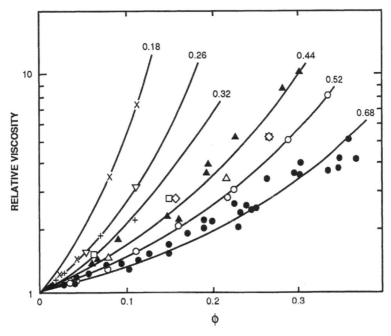

FIGURE 3.13-5. Relative viscosity-concentration behavior of various suspensions: solids in molten polymers. The numerical values for each curve denote the value of the parameter A of Equation (3.13-5). Legend is as follows: ● Glass spheres; ○ Natural calcium carbonate; △ Talc; ▲ Precipitated calcium carbonate; □ Glass fibers; aspect ratio = 6; ◇ Glass fibers; aspect ratio = 8; + Carbon fibers; aspect ratio = 18; ▽ Glass fibers; aspect ratio = 23; × Carbon fibers; aspect ratio = 27. Reproduced from Kitano, Kataoka, and Shirota [10] with permission.

that this line cannot be extrapolated unequivocally to smaller aspect ratios and probably cannot be extended to much larger ones.

There are very few literature data with which the applicability of Equation (3.13-5) to suspensions can be checked further. Lee et al. [17] report two measurements on SMC molding compounds but do not report the aspect ratio of the fibers in the sample; however, the concentration dependence of the reduced viscosity is much smaller than would be predicted from Equation (3.13-5) for a fixed value of A. In a second tabulation in their paper, no discernible variations in viscosity with fiber concentration were reported. This invariance would appear to be an unreasonable behavior in systems in which fibers were not degraded during processing. Maschmeyer and Hill [18] also report two data points, but the fiber aspect ratio is not given. The concentration dependence is, however, well portrayed by

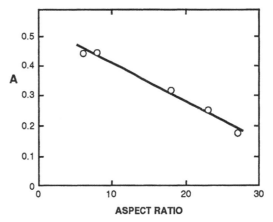

FIGURE 3.13-6. Dependence of the parameter A, Equation (3.13-5), upon aspect ratio of fibers suspended in molten polymers [10]. Reproduced with permission.

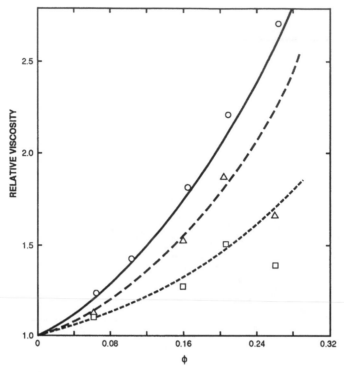

FIGURE 3.13-7. Relative viscosity-concentration curves for glass beads suspended in polypropylene. The three curves depict results at shear rates of zero (○), 100 (△), and 1000 (□) sec^{-1}. Reproduced from reference [20] with permission.

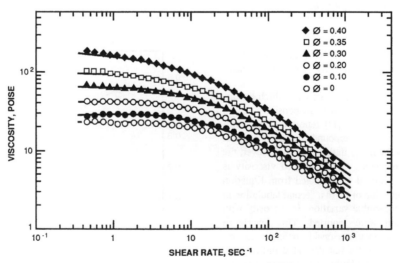

FIGURE 3.13-8. Viscosity-shear rate curves for glass beads suspended in polyisobutylene decalin solutions. Reproduced from Nicodemo, Nicolais, and Landel [21] with permission.

Equation (3.13-5), and the aspect ratio that one may back-calculate by fitting the data and using Figure 3.13-6 is not an unreasonable one.

There is a further set of data for particles of non-simple geometry that deserves mention. These data are reported in the classic and elegant study of Lewis and Nielsen [19], who assembled small glass beads into irregular agglomerates containing dimers, trimers, . . . up to agglomerates of as many as 250 particles. The agglomerates were formed by sintering, hence were permanent. Suspensions of each of the 10 classes of agglomerates were studied over a range of concentrations. The principal conclusion drawn was that the parameter K of Equation (3.13-4) increases with agglomerate irregularity and that Equation (3.13-4) correlates the data well, provided ϕ, the volume fraction, is that of the agglomerated spheres plus the liquid imbibed in the interstices of the permanent agglomerate. This conclusion is eminently reasonable, and since Equations (3.13-4) and (3.13-5) generally predict similar trends (see Figure 11 of reference [11]), it probably means that Equation (3.13-5) also correlates their data.

In summary, there exist too few good measurements of the viscosity of suspensions of irregular particles. However, all of them seem to be consistent with the one-parameter empirical equation:

$$\mu_r = \frac{1}{\left[1 - \dfrac{\phi}{A} \right]^2}$$

Therefore, its use is recommended generally, albeit with caution.

Effects of Increasing Deformation Rates and of Fluid Elasticity

In much of the work discussed thus far, care was taken to restrict measurements to low deformation rates so that shear-thinning effects—due either to alignment of suspended unsymmetric particles or to the shear-thinning behavior of a polymeric suspending liquid—did not complicate the observed behavior. We now wish to consider these effects of finite fluid deformation rate levels.

Capillary measurements of the power law exponents at shear rates in the range 50–600 sec^{-1} are reported for polypropylene melts filled with glass beads [20] as follows:

Volume Fraction Solids, ϕ	Power Law Flow Index, n
0	0.45
0.06	0.41
0.10	0.34
0.16	0.32
0.21	0.32
0.26	0.30

The increases in nonlinear flow behavior with increasing solids content presumably arise from alignment of groups or layers of the spherical beads. Since the viscosity decreases more rapidly with increasing shear rate for the most highly filled systems, the relative viscosity also decreases with increasing shear rate, as shown in Figure 3.13-7 from Faulkner and Schmidt, and over a wider concentration range for polymer solutions [21] in Figure 3.13-8.

Correlation of all of these viscosity data has been considered by many authors, but the simplest and evidently most definitive results are those of Kataoka and co-workers [11], Sundstrom [22], and Jarzebski [23]. Kataoka et al. [11] show that Equation (3.13-5) usually works well, at least at high deformation rate levels, *provided the relative viscosity* (i.e., the viscosity of the suspension divided by that of the carrier fluid) *is defined as that at the same shearing stress level.* Figure 3.13-9 shows their results. These conclusions, developed for particles suspended in molten polymers, were extended to polymer solutions by Sundstrom [22]. Both authors indicate some deviations from this simple behavior at the highest (40 volume percent) suspension concentrations, but in general the agreement is good. Jarzebski [23] recommends, instead, that one employ the theoretical Equation (3.13-2) developed by Frankel and Acrivos, and he has extended it to include shear-thinning, power law behavior. He has evaluated its use with the data of Nicodemo et al. [21] for suspensions in polymer solutions and of Chan et al. [24] for suspended solids in molten polymers; he finds acceptable agreement between prediction and experiment. One important result of the Jarzebski analysis [23] is to show that at high solids concentrations the effect of concentration upon the system viscosity decreases as the power law exponent decreases. This effect of concentration is to

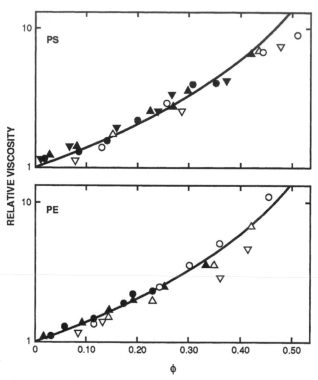

FIGURE 3.13-9. Relative viscosity-concentration behavior of suspensions of glass beads in molten polystyrene and molten polyethylene. Solid line depicts Equation (3.13-5) (with $A = 0.68$) with relative viscosities computed at equal values of the shearing stress. Reproduced from Kataoka, Kitano, Sasahara, and Nishijima [11] with permission.

enable some increase in the ratio ϕ/ϕ_m, i.e., in the solids loading, when one uses a more nonlinear suspending fluid.

Calculation of relative viscosities using Equations (3.13-2) and (3.13-5) shows, perhaps surprisingly, that over the concentration range, $0.2 \leq (\phi/\phi_m \text{ or } \phi/A) \leq 0.6$, the two equations predict values for the relative viscosities within 10%, and since most available data are restricted to this limited concentration range (as are most applications of interest), there is little basis at present for choosing one equation over the other.

It would, of course, be desirable to have available a data analysis that employs all of the literature measurements including the excellent data of Newman and Trementozzi [25], which extend to especially high concentration levels. Until this analysis is done, our recommendation is to prefer the theoretical analyses of Frankel and Acrivos [3] and Jarzebski [23] if one is dealing with suspensions of spheres or nearly spherical particles, and Equation (3.13-5) if one must deal with highly irregular or fibrous suspensions.

As one moves from low to high deformation rates, one must also determine how the *elastic properties* of the suspension (normal stresses, storage modulus) are dependent upon concentration of the suspended solids. There are only limited data with which to assess these effects and no evident theoretical calculations. Faulkner and Schmidt [20] show that, in capillary measurements, the Bagley end correction for the pressure drop required to force the fluid to flow into the tube is appreciably smaller for filled systems than it is for the polypropylene melt alone. Correspondingly, the die swell at the exit of the capillary was observed to decrease with increases in particle concentration in both this work and that of Newman and Trementozzi [25]. Nishimura and Kataoka [26] also report similar results for a variety of fillers in polypropylene and polystyrene melts. Measurements of the storage (elastic) and loss (viscous) moduli were correlated by Faulkner and Schmidt [20]:

$$\text{Storage:} \quad G_R' = 1.0 + 1.8\,\phi \qquad (3.13\text{-}6)$$

$$\text{Loss:} \quad G_R'' = 1.0 + 2.0\,\phi + 3.3\,\phi^2 \, (0 < \phi < 0.26)$$
$$(3.13\text{-}7)$$

In these equations the subscript R denotes "reduced," i.e., the property of the suspension divided by that of polymeric melt. Equations (3.13-6) and (3.13-7) show that the force required to overcome elastic and viscous resistances to deformation increases with solids content, but the viscous effects as measured by G_R'' are appreciably greater. Measurements by Kitano, Kataoka, and Magatsuka [27] extend the results of Faulkner and Schmidt and are consistent with them. In other words, the comparative importance of the elastic forces decreases as the solids concentration levels increase. This is a comforting result in the sense that, if one knows a molding operation to be independent of elastic fluid properties when one is dealing with the neat polymer (as is frequently the case), then this result will surely be so for suspensions. The only other measurements of normal stresses are for fiber suspensions [24] that appear to show the inverse of this behavior in that the normal stresses rise slightly more rapidly than does the viscosity ratio upon addition of solids to molten polymers. However, this effect may be an artifact related to the tendency of suspended fibers to adhere to one another and to "rope" around the center for rotational instruments, as described first by Nawab and Mason [28].

In summary, as one moves into the shear-thinning range of deformation rates, one finds that the addition of suspended solids increases the shear-thinning. It appears that this viscous behavior can be predicted correctly using either Jarzebski's theoretical analysis [23] or the empirical equation of Kataoka et al. [11] and Sundstrom [22]; both approaches predict a similar behavior in the range of particle concentrations studied to date.

Limited data on elastic fluid properties generally suggest that the relative importance of fluid elasticity decreases with increasing solids content of the suspension.

Suspensions of Long Fibers

Except for the last paragraphs the previous discussion was restricted to symmetric or nearly symmetric particles (aspect ratio ≈ 1.0) and to short fibers (aspect ratio as great as 27).

There are very few published measurements available of the rheological properties of fiber suspensions in shearing flows at the concentration levels of interest in practice and only one of these contributions, Chan, White, and Oyangi [24], employed polymeric fluids for the suspending liquids. Their fibers had aspect ratios of 80 ($\phi = 0.19$ suspension) and 160 ($\phi = 0.08$ suspension). Figures 3.13-10 and 3.13-11, taken from their work, summarize data in the literature as well as their own measurements. We should like to make several comments about these results:

1. All of the curves shown in Figure 3.13-10 fall *enormously* lower than those of Figure 3.13-5 for an aspect ratio of 27. The primary reason is that the viscosity data of Chan et al. [24] refer to a shear rate of 10 sec⁻¹, well removed from the low shear rate conditions of Figure 3.13-5. However, even the zero shear rate results of Chan and co-workers seem

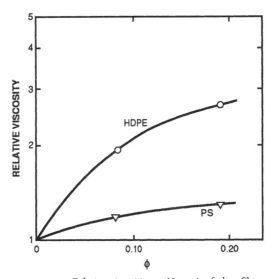

FIGURE 3.13-10. Relative viscosities at 10 sec⁻¹, of glass fibers suspended in molten polymers. Reproduced from Chan, White, and Oyanagi [24] with permission of the copyright owner, John Wiley & Sons, New York.

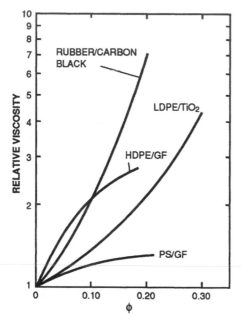

FIGURE 3.13-11. Comparison of the relative viscosities of suspensions of rubber/carbon black, high density polyethylene/glass fibers, low density polyethylene/TiO₂ and polystyrene/glass fibers [24]. Reproduced with permission.

more characteristic of an aspect ratio of 15–20 instead of their much larger values. The reasons for this discrepancy are not clear.

2. The zero shear rate data for the polystyrene (PS) suspension are about the same as those for HDPE. That is, the lower position of the PS data in Figure 3.13-10 is at least partly an artifact due to comparison of viscosities at equal shear rates (10 sec⁻¹) instead of at equal stress levels and the very great shear-thinning of the PS. Whether a comparison at equal stress levels or the use of the Jarzebski equations would fully resolve these differences is not known.

3. The concave-downward shape of the curves for the HDPE and PS suspensions in Figures 3.13-10 and 3.13-11 is almost certainly due to the decrease in the fiber aspect ratio from 160 to 80 as the concentration was increased from 8 to 19%. Both curves, if the aspect ratio of the suspended fibers were constant, would be expected to be concave-upward (as in Figure 3.13-5).

4. At small shear rates the data for the PS and HDPE suspensions in Figure 3.13-11 would fall above the curve labelled "LDPE-TiO₂", as expected from Figure 3.13-5. The LDPE-TiO₂ curve falls only slightly above that in our Figure 3.13-1 and thus is roughly in agreement with it. The slightly greater relative viscosities of the LDPE-TiO₂ system could reflect a (very modest) adsorption of the polymer onto the particles, thus effectively increasing their size and volume fraction ϕ to greater values than those calculated from the mixture composition. In the case of the carbon black–rubber system, much adsorption of polymer onto the particles is expected, thus greatly increasing ϕ above its computed value. (We shall consider adsorption effects somewhat further in subsection 3.13.3, where explicit data are available.) Additionally, some authors note that carbon black may bond chemically with free radicals on the polymer molecule formed during milling operations to produce new networks. In other words, the curve for carbon black suspensions may be regarded as anomalous. White and Crowder [29] provide a good review of the peculiar structural nature of carbon black–rubber systems.

Miles, Murty, and Modlen [30] have made capillary viscometry studies of dilute fiber suspensions. They note that long fibers in small tubes must be completely aligned, and, under these circumstances, the suspension viscosity should be independent of the fiber aspect ratio, and this was found to be the case. However, the entry pressure loss, associated with disruption of entangled fibers on flowing into the capillary, is large and increases rapidly with fiber L/D and the volumetric concentration ϕ. (Their concentration range was from 0 to 3% by volume, and the fiber aspect ratios were approximately 1, 120, and 500.) Dinh and Armstrong [31] computed the ratio of stresses required to deform such a random suspension of fibers, as compared to the stresses in the carrier liquid:

$$\frac{\mu_a}{\mu} \approx \phi \left(\frac{L}{D} \right)^2 \qquad (3.13\text{-}8)$$

Horie and Pinder [32] studied flow of fibers suspended in low viscosity liquids employing a Couette (rotating concentric cylinder) viscometer. They note that the suspensions did not remain homogeneous in fiber conformation when shearing began: in some devices, perhaps all, the suspension micro-structure changes and it becomes meaningless to talk of a uniform "viscosity" for the whole system. Many of these extreme complications should largely disappear when the viscosity of the suspending liquid becomes large enough, as in the work of Chan et al. [24].

Wissbrun [33] has noted that concentrated suspensions of rigid fibers might exhibit rheological properties comparable to those of polymeric liquid crystalline fluids, for which the "rigid rods" model is used in the derivation of the flow properties. In such predictions the viscosity at low concentration levels is expected to increase rapidly with increasing concentration of the particles because of the rapidly increasing interactions between particles as they become crowded more closely together. However, at a "critical" concentration level, random packing ceases to be possible and further increases in particle concentration imply a progressively more orderly anisotropic "liquid crystalline" structure of the rods in the suspension; these rods may now slide readily past one another. Hence, above the critical concentration level, further increases in particle concentration imply progressive *decreases* in the viscosity of the system, at least until very high concentration levels are reached. To our knowledge, this behavior has not been observed in macroscopic suspensions, but it is reasonable to look for this phenomenon.

In summary, suspensions of long fibers in shearing flows may exhibit very high stress levels during transient shearing deformations in which fiber alignment occurs. There are few quantitative data with which to verify this prediction, but the problem may be of importance in flow into dies or through gates. Once a steady state degree of alignment has been achieved in viscous fluids, the system behavior does not appear to be especially complex, but the available data are too limited to enable definitive design equations to be recommended.

Elongational Flows

Several authors have predicted increases in the resistance to deformation when elongational flows, rather than shearing deformations, are imposed on a suspension [34–38]. In the case of spherical particles the differences are, however, only of modest magnitude. As seen in Figure 3.13-2, which reaches to extraordinarily high concentration levels for a unimodal particle size distribution, the orifice viscometer (in which the defor-

mation process is largely extensional) yields results that, as a first approximation, are very comparable to those obtained in the two shearing instruments. Patzold [7] shows data for a second suspension in which the differences are somewhat larger, but again they do not differ dramatically.

In marked contrast to this simple behavior of spherical particles in a suspension, fibrous suspensions show enormously greater resistances to deformation in extensional flows. The Batchelor [35] equation

$$\lambda = \mu \left[3 + \frac{4\phi(L/D)^2}{3 \ln (\pi/\phi)} \right] \qquad (3.13\text{-}9)$$

in which λ, an "extensional viscosity" or ratio of stress to deformation rate, contains the Newtonian term 3μ to which is added the contribution of the suspended fibers. It is restricted to a semiconcentrated range of fiber concentrations such that the distance H between the (aligned) fibers is greater than the fiber diameter but smaller than the fiber length:

$$L \gg H \gg D$$

As such, it approaches but does not cover the range of interest in processing short-fiber composites, and its extension to higher concentration levels, theoretically as well as experimentally, is of interest. Presumably, the volumetric concentration term will become replaced by a dependence upon some function of ϕ/ϕ_m, as in Equations (3.13-2) and (3.13-5).

As a fluid filament containing aligned fibers is stretched, the liquid between the fibers must be expelled to enable them to approach each other more closely. Doubling the length of the fibers doubles the volume of fluid to be expelled between a group of fibers, and it must be expelled over twice the distance; hence, there arises a quadratic dependence upon L/D that endows Equation (3.13-9) with extraordinarily high levels of the predicted value of λ. Figure 3.13-12 illustrates this exceptional increase in resistance to deformation upon addition of only 1/10%–1% of fibers to the suspending liquid; Figure 3.13-13 depicts the extraordinary agreement between Equation (3.13-9) and the data. As great as these resistances to extensional deformation may be, we should note that the stresses predicted by the Dinh-Armstrong analysis [31] in Equation (3.13-8), for disentanglement of disoriented fibers by a shearing flow are still greater than those of Equation

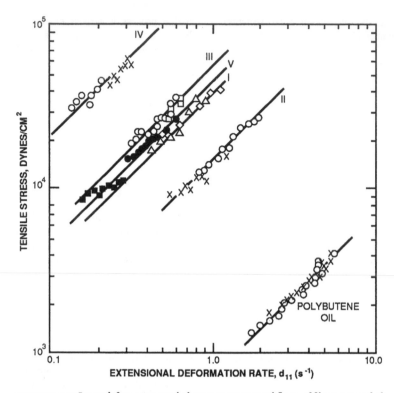

FIGURE 3.13-12. Stress-deformation rate behavior, in extensional flows, of fibers suspended in a Newtonian oil. The fiber L/D values and volumetric concentrations employed were as follows:

Curve		L/D
I	0.0093	282
II	0.00099	586
III	0.00287	586
IV	0.0089	586
V	0.00096	1259

Differing symbols on a given curve denote different experiments using the same fluid. Reproduced from Mewis and Metzner [39] with permission of Cambridge University Press.

(3.13-9): for a 1% suspension of fibers having (L/D) of 100, the calculated ratios of suspension to solution deformational resistances are 100 and 3.6, respectively.

Figure 3.13-14 shows the additional data available for elongational flows of fiber suspensions [40,41]. Although agreement with the Batchelor theory does not appear to be exact, it is good and the major trends are given well.

If viscoelastic and shear-thinning molten polymers are employed as the suspending liquid for the fiber suspension, μ, the viscosity of the suspending liquid should be evaluated at the high shear rates extant between the particles. These shear rates can be estimated, at least roughly, using lubrication theory [42,43]. There are two analyses of the effect of viscoelastic fluid properties upon the stresses required to deform viscoelastic suspensions [43,44]; they do not even agree as to the direction of the change to be expected when a viscoelastic carrier liquid replaces a Newtonian fluid; hence, definitive conclusions are not yet possible. Likewise, very few data are available on this subject; Chan et al. [24] summarize what there is.

In summary, the inclusion of fibrous particles in a viscous Newtonian liquid leads to very great increases in the resistance to extensional deformations, even at the very low solids concentrations studied to date. These effects are well documented analytically and experimentally. Very little work of either kind exists for concentrated suspensions or for suspensions in viscoelastic fluids.

If the high resistance to extensional deformations noted above were also to occur in concentrated fiber-polymer mixtures, an appreciable modification of the traditional "fountain flow" kinematics of the fluid at the advancing front [45,46] during mold-filling operations might be expected, with the result that the fluid and fiber orientation in the filled mold may differ appreciably from those found to occur with homogeneous, single phase liquids. Studies of the differences in flow patterns between fiber filled polymers and neat resin have indeed begun [47,48]. It is not obvious that they confirm such predictions, although parts molded of the fiber filled fluid did, by warpage, reveal larger residual stress levels, indicating, in general, a greater previous resistance to deformation that may be a fruitful area for further study.

3.13.3 Suspensions in Liquids of Low Viscosity

The description in the previous subsection was largely devoted to idealized systems in the sense that interparticle forces could be neglected; i.e., they may be swamped, as it were, by the high level of the viscous stresses generated in the deforming fluid-solids mixture. Many processing problems will fall into such a regime, and since these idealized systems are the best understood ones, the emphasis on them is also desirable from the viewpoint of establishing a firm foundation for our general understanding. What complications arise, however, as the ratio of interparticle attractions to viscous forces increases? We will not treat this subject in detail here but do wish to illustrate the possible complications to which one needs to be alert.

Figure 3.13-15 depicts viscosity measurements for suspensions of TiO_2 particles in high density polyethylene melts [49]: where HDPE is not an especially low viscosity fluid, the interparticle forces are substantial in comparison to viscous stresses. We see that, even

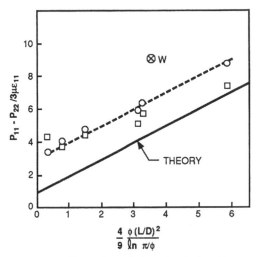

FIGURE 3.13-13. Comparison of elongational viscosities predicted from Equation (3.13-9) with data from Figure 3.13-12 and from Weinberger and Goddard [40]. Reproduced from reference [39] with permission.

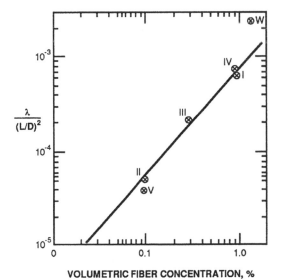

FIGURE 3.13-14. Normal stresses (extensional viscosities) for fiber suspensions. Symbols denote different velocities in the orifice device used; W denotes data from Weinberger and Goddard [40]. Figure reproduced from Kizior and Seyer [41] with permission.

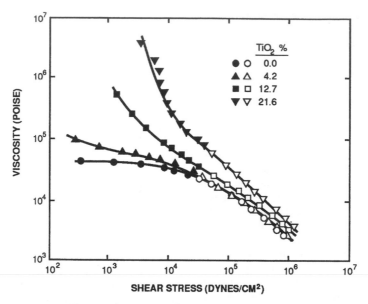

FIGURE 3.13-15. Viscosity-shearing stress behavior of high density polyethylene melts containing TiO_2 particles. Note the progressive development of a yield stress (curves concave upward at low stress levels) as TiO_2 concentration increases. Concentrations are in volume %. Reproduced with permission from Minagawa and White [49].

at the very low solids concentration of 4%, there appears to be some particle-particle interaction (perhaps through adsorbed polymer) to cause the viscosity-stress curve to bend upward as progressively lower stresses (or deformation rates) are studied. At the highest solids concentration, an apparent yield stress of the order of 10^3 dynes/cm² must be exceeded for flow to occur. As the stress level in a fluid sample is increased to higher values, the "structure" responsible for the yield stress rapidly crumbles, and at high stress or deformation rate levels, the viscosity of the suspension is only marginally greater than that of the polymer alone. Changes in the particle surface, whether they are intentional or arise out of subtle changes in the manufacturing process, have large effects on the rheology of these suspensions [50]. In aqueous systems the particle-particle interaction through electrostatic forces is enormously greater, and even very dilute suspensions may require that a yield value be exceeded before flow can be initiated. Figure 3.13-16 shows dramatically how a slurry of low concentration (7% by volume) may resist deformation until the imposed stress exceeds ca. 200–250

dynes/cm², and how a three-fold further increase in stress level results in more than a two order of magnitude increase in deformation rate. As expected, these electrostatic forces are very sensitive to ionic additives or dispersants; "process aids," for solids dispersed in molten polymers, may also enable large changes in fluid viscosity at low stress levels, though perhaps not as dramatic as those of Figure 3.13-16.

Simple effects, due to particle wettability and solvation that may change a suspension from one of inert, noninteracting particles (as in the case of dilute suspensions of sand in water) to one of "sticky" or adhering particle networks, have been vividly illustrated in classic papers by Verway and DeBoer [52] and Pryce-Jones [53]. Their examples are as follows: in one instance, a dilute and very fluid coal-in-water slurry (in every way comparable to a dilute suspension of sand in water) is converted into a solid of substantial rigidity by addition of only a few drops of CCl_4 per deciliter of suspension. In a second example, starch particles suspended in an inert organic liquid likewise develop a rigid network upon addition of a few drops of water.

We have alluded to effects of polymer adsorption onto particles in a number of instances, and several good sets of measurements of the magnitude of this phenomenon are available [54,55]. This adsorption, in an extreme case like that of the carbon black–rubber system discussed earlier, may provide cross-links between polymer chains with inordinate influences on the flow behavior of the system. However, even in the simplest possible case in which polymer is adsorbed from solution to leave noninteracting suspended particles, the effects may be major ones. Mewis [54] and Cohen and Metzner [55] show that adsorbed films having a thickness of about 1 μm commonly are adsorbed onto the surfaces of a wide variety of solids from a similar variety of polymer solutions. If the particle concentration is initially high and the particles themselves are in or near a μm size range, the change in the effective size and, hence, concentration of the particles may be major and cause an effective translation to the right along the horizontal axis of curves such as those in Figures 3.13-1 and 3.13-2, with a large increase in viscosity resulting.

If a suspension of particles at rest is sufficiently concentrated to produce an approach to a closely packed crystalline array, any attempt to shear the suspension must require one particle to move up and over its nearest neighbors, since it cannot merely slide past them as in the example of a dilute system. Such a relative particle motion may require an expansion or volumetric dilation of the system. When such a volume change occurs, there may now be insufficient liquid phase remaining in the suspension to lubricate this motion, implying a rise in the system viscosity that may accelerate as the deformation rate process increases and progressively larger fractions of all the solids come into relative motion simultaneously. Figure 3.13-17 shows a gentle, mild form of this rheological dilatancy. Figure 3.13-18 shows how this rheological dilatancy may develop progressively as the concentration of suspended solids in a fluid increases. At the 12% concentration level, the rheological behavior is a simple Newtonian one, and all complications are absent. At the 27% concentration level, the fluid exhibits an apparent yield value and is, consequently, very shear-thinning at low deformation rates; at 200–300 sec⁻¹ it is essentially Newtonian, and at higher deformation rates it becomes gently shear-thickening or rheologically dilatant, much as in the case of the glass bead suspension of Figure 3.13-17. Further increases in solids concentration exacerbate these trends until, at 47% solids, it appears

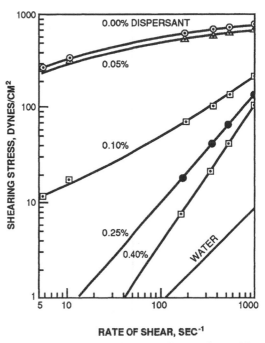

FIGURE 3.13-16. Rheological behavior of a 7% slurry of hematite in water, illustrating dramatic effects of addition of small concentrations of a lignosulfonate dispersant. Reproduced from Jennings [51] with permission.

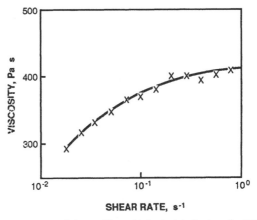

FIGURE 3.13-17. Dilatant (shear-thickening) behavior of a 54 volume % suspension of glass spheres in oil. Reproduced from Patzold [7] with permission.

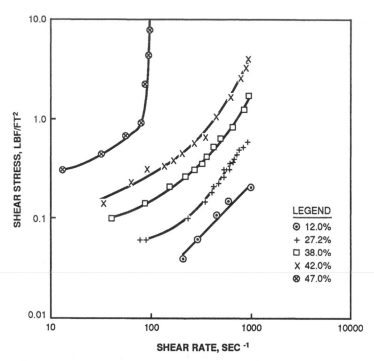

FIGURE 3.13-18. Development of dilatancy in aqueous TiO_2 suspensions; concentrations are in volume %. Reproduced from reference [56] with permission of the copyright owner, John Wiley & Sons, New York.

impossible to deform the system at shear rates in excess of 100 sec^{-1}. In Figure 3.13-19 one sees that the forced imposition of high deformation rates upon highly loaded systems may lead to order of magnitude discontinuities in the rheological behavior; Hoffman [57] has shown beautifully how an ordered, hexagonal structure of the solid particles in the suspension at rest or undergoing low deformation rates changes to a disordered (but still deformable) array at deformation rates just beyond the discontinuity.

In summary, we see that particle-particle associations may lead to development of high yield values in concentrated suspensions, that surface treatment of the particles may be used to reduce or eliminate this behavior, and that these particle associations may be electrostatic in origin or due to particle solvation (stickiness). (Mechanical entanglement in dilute suspensions of fibrous particles is also known to lead to yield values.) Our emphasis upon the question of whether yield values occur arises for two reasons. First, there has been a rather general misunderstanding of the influence of yield values upon the velocity field of a deforming material: yielding and flow must usually occur everywhere in flows through complex geometries [58] which may imply unexpectedly high stress levels. Second, the impossibility of removing small bubbles from fluids exhibiting yield values may lead to significant defects in fabricated parts produced from polymeric systems with a yield value.

Adsorption of polymers onto the surface of particles appears to be ubiquitous and influences suspension rheology. Highly concentrated systems at high deformation rates generally become dilatant (shear-thickening), which may impose severe restrictions on the processability of such suspensions.

Because these "complications" have not been dealt with in detail here, the interested reader should look to other treatises [59,60] for information on this subject.

Finally, the author would like to draw attention to the excellent overall review of this subject published by Jeffrey and Acrivos [61].

3.13.4 Suspension Inhomogeneity and Ancillary Complications

An implicit assumption in the previous discussion was that we were treating fluids of homogeneous structure and composition. As the concentration of suspended solids increases towards its maximum possible value, this condition becomes progressively more difficult to achieve, even if the size of the suspended particles is small in comparison to the apparatus or system dimensions and filtration does not occur.

During Poiseuille flow through a duct, any particles suspended in a fluid move radially inward, away from the wall [62,63], a phenomenon termed the *Segre-Silberberg effect*. The effects of this phenomenon on injection molding processes may be of surprising magni-

tude, as shown in a pair of classic papers by Kubat and associates [64,65]. Consider flow into a long empty tube. As the tube is being filled, the suspended particles migrate from the wall towards the centerline, thereby entering the faster moving stream and being convected downstream more rapidly than the average of the suspending liquid. When the tube or mold is filled, there is consequently an axial solids concentration gradient. The effect was observed to be greatest for the largest particles [65]; consequently, a mixture of particle sizes is also separated. The ratio of the concentration found at the downstream end of a 2 m long tube to that at the inlet was as great as 6:1! Similarly, impressively documented effects of the Segre-Silberberg phenomenon in nonisothermal flows have been presented in a series of papers by Schmidt [66–68]. In this case the fluid velocity field is quite complex, and our quantitative understanding is not yet complete, but especially disturbing is the experimental observation of extensive resin-rich layers that appear to be nearly devoid of the suspended particles after a center-gated plaque was

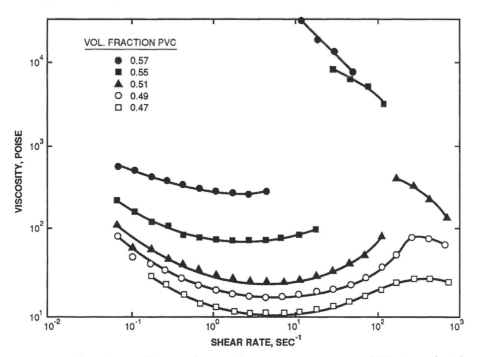

FIGURE 3.13-19. Shear-thinning, dilatancy and viscosity discontinuities in suspensions of 1.25 micron polyvinyl chloride particles in dioctyl phthalate. Reproduced from Hoffman [57] with permission of the copyright owner, John Wiley & Sons, New York.

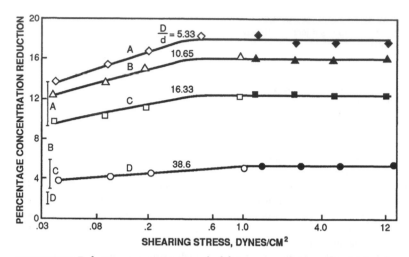

FIGURE 3.13-20. Reductions in concentration of solids in a suspension upon flow into a tube from an upstream reservoir. Neutrally buoyant, rigid spheres with initial concentration of 23.5 % by volume. D/d denotes the ratio of tube diameter to diameter of the suspended particle. Suspending liquid viscosity was 1.8 cp. (open symbols) and 23.5 cp. (solid symbols). Reproduced from Seshadri and Sutera [71] with permission of the copyright owner, John Wiley & Sons, New York.

molded. An initial study of co-injection of reinforced polymers, in which all of these complications are also present, has been presented by Akay [69].

Consider as a further example flow from a reservoir or chamber of large diameter into a small tube. When pressure is applied upstream to initiate the flow process, it is possible that a suspension moves forward homogeneously. But it is also possible that, in concentrated suspensions with associating or at least non-independent particles/fibers [see Figures 3.13-15 and 3.13-18 and Equation (3.13-8)], the "mat" of solids may simply compress, expelling the suspending liquid alone into the downstream tube. Since even small changes in concentration, at high concentration levels, lead to further increases in the resistance to flow of that part of the system containing the solids, there may be an exacerbating feedback leading to complete exclusion of the particles from the downstream flow [9,70,71]. However, even in suspensions of modest concentration level, in which blockage does not occur, there may be substantial concentration reductions on flow into the smaller tube, as shown in Figure 3.13-20.

The nature of the solids in concentrated closely packed suspensions has been alluded to on a number of

occasions, but we have not quantified that description. Reference to solid-state crystal structures shows that several alternate structures are possible, in which the elements are closely packed. While a concentrated suspension is being sheared, it seems likely that transitions from one structure to another could take place, resulting in progressive changes in rheological properties [7,9]. These transitions may be one of the reasons for many of the blips, spikes, and other aberrations in rheological measurements frequently reported in the literature [9], for some of the changes in viscosity noted over long periods of shearing time, and for the appreciable differences between the curves of Figures 3.13-1, 3.13-2, and 3.13-3.

During rheological measurements, apparent "slip" phenomena frequently arise out of the fact that the solids packing density in a suspension must be lower at a wall than in the interior of a suspension [6,9]. Well developed means for correcting for this lacking density are available [72,73]; its importance in processing of suspensions has not been studied extensively to our knowledge, but references [74–76] summarize the available literature.

Flows of molten polymers and of some solutions are

known to become unsteady at sufficiently high flow rates, especially in the case of flow into abrupt contractions [76–79]. This "melt fracture" phenomenon is appreciably modified in the case of suspensions due to the influence of the suspended particles. An excellent study of the subject has been presented [80], although general conclusions do not yet appear to be possible.

3.13.5 References

1. EINSTEIN, A. *Ann. Phys.*, 19:289 (1906) and 34:591 (1911).
2. HAPPEL, J. and H. Brenner. *Low Reynolds Number Hydrodynamics*. Prentice-Hall, Englewood Cliffs, NJ (1965).
3. FRANKEL, N. A. and A. Acrivos. *Chem. Eng. Science*, 22: 847–853 (1967).
4. MOONEY, M. *J. Colloid Sci.*, 6:162 (1951).
5. THOMAS, D. G. *J. Colloid Sci.*, 20:267–277 (1965).
6. COHEN, Y. and A. B. Metzner. *AIChE J.*, 27:705–715 (1981).
7. PATZOLD, R. *Rheol. Acta*, 19:322–344 (1980).
8. BRADY, J. F. and G. Bossis. In *Advances in Rheology, Vol. 2.* B. Mena, A. Garcia-Rejon and C. Rangel-Nafaile, eds., Pub. by Universidad Nacional Autonoma de Mexico, pp. 475–480 (1984).
9. CHENG, D. C.-H. *Powder Technology*, 37:255–273 (1984).
10. KITANO, T., T. Kataoka and T. Shirota. *Rheol. Acta*, 20: 207–209 (1981).
11. KATAOKA, T., T. Kitano, M. Sasahara and K. Nishijima. *Rheol. Acta*, 17:149–155 (1978).
12. DE SOUSA, J. A. In *Advances in Rheology, Vol. 3.* B. Mena, A. Garcia-Rejon and C. Rangel-Nafaile, eds., Pub. by Universidad Nacional Autonoma de Mexico, pp. 439–446 (1984).
13. EVESON, G. F. *Rheology of Dispersed Systems*. Pergamon, London (1959).
14. FARRIS, R. J. *Trans. Soc. Rheol.*, 12:281–301 (1968).
15. HENDERSON, C. B., R. S. Scheffee and E. T. McHale. Manuscript prepared for Cleveland AIChE Meeting (August 1982).
16. CHONG, J. S., E. B. Christiansen and A. D. Baer. *J. App. Poly. Sci.*, 15:2007–2021 (1971).
17. LEE, L. J., L. F. Marker and R. M. Griffith. *Polymer Composites*, 2:209–218 (1981).
18. MASCHMEYER, R. O. and C. T. Hill. *Trans. Soc. Rheol.*, 21:195–206 (1977).
19. LEWIS, T. B. and L. E. Nielsen. *Trans. Soc. Rheol.*, 12: 421–443 (1968).
20. FAULKNER, D. L. and L. R. Schmidt. *Polymer Eng. Sci.*, 17:657–665 (1977).
21. NICODEMO, L., L. Nicolais and R. F. Landel. *Chem. Eng. Science*, 29:729–735 (1974).
22. SUNDSTROM, D. W. *Rheol. Acta*, 22:420–423 (1983).
23. JARZEBSKI, G. J. *Rheol. Acta*, 20:280–287 (1981).
24. CHAN, Y., J. L. White and Y. Oyanagi. *J. Rheol.*, 22:507–524 (1978).
25. NEWMAN, S. and Q. A. Trementozzi. *J. Appl. Polym. Sci.*, 9:3071–3089 (1965).
26. NISHIMURA, T. and T. Kataoka. *Rheol. Acta*, 23:401–407 (1984).
27. KITANO, T., T. Kataoka and Y. Magatsuka. *Rheol. Acta*, 23:408–416 (1984).
28. NAWAB, M. A. and S. G. Mason. *J. Phys. Chem.*, 62: 1248–1253 (1958).
29. WHITE, J. L. and J. W. Crowder. *J. Appl. Poly. Sci.*, 18:1013–1038 (1974).
30. MILES, J. N., N. K. Murty and G. F. Modlen. *Polym. Eng. Sci.*, 21:1171–1172 (1981).
31. DINH, S. M. and R. C. Armstrong. *J. Rheol.*, 28:207–227 (1984).
32. HORIE, M. and K. L. Pinder. *Can. J. Chem. Eng.*, 57:125–134 (1979).
33. WISSBRUN, K. F. Private communication (1984).
34. GIESEKUS, H. *Rheol. Acta*, 2:122–130 (1962).
35. BATCHELOR, G. K. *J. Fluid Mech.*, 44:419–440 (1970).
36. BATCHELOR, G. K. *ibid.*, 46:813–829 (1971).
37. BATCHELOR, G. K. and J. T. Green. *ibid.*, 56:401–427 (1972).
38. BATCHELOR, G. K. *ibid.*, 83:97–117 (1977).
39. MEWIS, J. and A. B. Metzner. *ibid.*, 62:593–600 (1974).
40. WEINBERGER, C. B. and J. D. Goddard. *Int. J. Multiphase Flow*, 1:465–486 (1974).
41. KIZIOR, T. E. and F. A. Seyer. *Trans. Soc. Rheology*, 18: 271–285 (1974).
42. DENN, M. M. *Process Fluid Mechanics*. Chapter 13, Prentice-Hall, Englewood Cliffs, NJ (1980).
43. METZNER, A. B. *Phys. Fluids*, 20:S145–S149 (1977).
44. GODDARD, J. D. *J. Non Newtonian Fluid Mech.*, 1:1–17 (1976).
45. TADMOR, Z. *J. Appl. Poly. Sci.*, 18:1753–1772 (1974).
46. ROSE, W. *Nature*, 191:242–243 (1961).
47. SCHMIDT, L. R. *J. Rheol.*, 22:571–588 (1978).
48. SCHMIDT, L. R. *Proc. 2nd World Congress of Chem. Eng.*, Vol. 6, pp. 516–518 (1981).
49. MINAGAWA, N. and J. L. White. *Poly. Eng. Sci.*, 15:825–830 (1975).
50. MINAGAWA, N. and J. L. White. *J. Appl. Poly. Sci.*, 20: 501–523 (1976).
51. JENNINGS, H. Y. *J. Am. Oil Chem. Soc.*, 46:642–644 (1969).
52. VERWAY, E. J. W. and J. DeBoer. *Rec. Trav. Chim.*, 57: 383–389 (1939).
53. PRYCE-JONES, J. *Proc. Durham Univ. Phil. Soc.*, 10:427–467 (1948).
54. MEWIS, J. "Rheology and Microstructure of Suspensions,"

Manuscript available from Katholieke Universiteit Leuven, Belgium.

55. COHEN, Y. and A. B. Metzner. *Macromolecules*, 15:1425–1429 (1982).

56. METZNER, A. B. and M. Whitlock. *Trans. Soc. Rheology*, 2:239–254 (1958).

57. HOFFMAN, R. L. *ibid.*, 16:155–173 (1972).

58. LIPSCOMB, G. G. and M. M. Denn. *J. Non Newtonian Fluid Mech.*, 14:337–346 (1984).

59. RUSSEL, W. B. "Effects of Interactions between Particles on the Rheology of Dispersions," *Theory of Dispersed Multiphase Flow*, R. E. Myer, ed. Academic Press (1983). See also *J. Rheology*, 24:287–317 (1980).

60. OTSUBO, Y. and K. Umeya. *J. Rheology*, 28:95–108 (1984).

61. JEFFREY, D. J. and A. Acrivos. *AIChE J.*, 22:417–432 (1976).

62. SEGRE, G. and A. Silberberg. *Nature*, 189:209 (1961) and *J. Fluid Mech.*, 14:115 (1962).

63. DENSON, C. D., E. B. Christiansen and D. L. Salt. *AIChE J.*, 12:589–595 (1966).

64. KUBAT, J. and A. Szalanczi. *Polym. Eng. Sci.*, 14:873–877 (1974).

65. BOROCZ, L. and J. Kubat. *Plastics and Rubber Processing*, pp. 82–86 (1979).

66. SCHMIDT, L. R. *Advances in Chemistry*, N. A. J. Platzer, ed., 142:415–429 (1975).

67. SCHMIDT, L. R. *Polym. Eng. Sci.*, 17:666–670 (1977).

68. SCHMIDT, L. R. *Science and Technology of Polymer Processing*, N. P. Suh and N. H. Sung, eds., MIT Press, pp. 315–328 (1979). See also U.S. Patent 4,029,841 (1977).

69. AKAY, G. *Polymer Composites*, 4:256–264 (1983).

70. RIGGS, L. C. B.Ch.E. thesis, University of Delaware (1956).

71. SESHADRI, V. and S. P. Sutera. *Trans. Soc. Rheol.*, 14:351–373 (1970).

72. MOONEY, M. *J. Rheol.*, 2:210 (1931).

73. COHEN, Y. and A. B. Metzner. *Chem. Eng. Progress Symposium Series*, 78(212):77–85 (1982).

74. MARKER, L. and B. Ford. *Proc. 32nd Annual Conf., Soc. Plastics Industry* (1977).

75. LEE, S. J., M. M. Denn, M. J. Crochet and A. B. Metzner. *J. Non Newtonian Fluid Mech.*, 10:3–30 (1982).

76. WINDHAB, E. and W. Gleissle. *Advances in Rheology, Vol. 2*. B. Mena, A. Garcia-Rejon and C. Rangel-Nafaile, eds., Pub. by Universidad Nacional Autonoma de Mexico, pp. 557–564 (1984).

77. SPENCER, R. S. and R. E. Dillon. *J. Colloid Sci.*, 4:241–255 (1949).

78. TORDELLA, J. P. *J. App. Phys.*, 27:454–458 (1956).

79. BOGER, D. V. *Advances in Transport Processes, Vol. 2*. A. S. Mujumdar and R. A. Mashelkar, eds., Wiley International (1982).

80. AKAY, G. *J. Non Newtonian Fluid Mech.*, 13:309–323 (1983).

Milton Keynes UK
Ingram Content Group UK Ltd.
UKHW020822141024
449569UK00008B/514